green guide

FROGS

....

OF AUSTRALIA

Gerry Swan

Series Editor: Louise Egerton

NEW
HOLLAND

First published in 2001 by
New Holland Publishers (Australia) Pty Ltd
Sydney • London • Cape Town • Auckland

14 Aquatic Drive, Frenchs Forest NSW 2086 Australia
24 Nutford Place, London W1H 5DQ United Kingdom
80 McKenzie Street, Cape Town 8001 South Africa
218 Lake Road, Northcote Auckland New Zealand

Series Editor: Louise Egerton
Project Manager: Rosemary Milburn
Editor: Maggie Aldhamland
Design and Cartography: Jenny Mansfield
Picture Research: Kirsti Wright
Illustrations: Gerry Swan and others as credited
Reproduction by: Pica, Singapore
Printed and bound by: Kyodo Printing

National Library of Australia Cataloguing-in-Publication Data
    Swan, Gerry.
    Frogs of Australia.
    Includes index.

    ISBN 1 86436 333 9

    1. Frogs — Australia. 2. Frogs — Australia — Identification.
    I. Egerton, Louise. II Title. (Series: Green guide (New Holland)).

    597.90994

The body copy is set in 9pt Cheltenham Light.

Photographic and Other Acknowledgments
Abbreviations: ANT = A.N.T. Photo Library; PRTF = Peter Rankin Trust Fund
Photographic positions: t = top, b = bottom, c = centre, m = main, i = inset, l = left, r = right, fc =
front cover, bc = back cover, ff = front flap
Ken Griffith: fc br, bc b, p. 6b, 9b, 16t, 17, 18t, 22b, 25t, 26t, 28, 30-31, 35t, 36t, 37t, 41i, 44t, 50-51,
52t, 53t, 54i, 57t, 62t, 67, 73t, 75t, 76t, 88t, 89t, 90t, 91b; **Pavel German:** fc tl, bc t, p. 3, 18i, 24t,
33t, 35i, 40, 48b, 53b, 55, 58t, 73i, 88b, 89b, 90b, 91t, 93, 94, 95 lt&b; **Geoff Swan:** fc l, p. 4b, 6t, 7t,
8, 9t, 1 1i, 12, 13b, 20b, 21b, 23t, 24b, 37b, 38b, 42t, 45b, 49t, 54t, 63t, 66, 70-72, 76b, 77, 83, 92,
95bl; **Michael Mahony:** fc c, ff, p. 4t, 10-11, 13t, 14b, 15, 20, 23b, 26b, 27, 29, 31i, 32, 34, 36b, 38t,
41t, 42b, 43t, 45t, 47, 48t, 56, 58b, 59t, 60-61, 62b, 64, 65b, 68-69, 74, 75b, 78-79, 84b, 85b, 87; **Mike
Anthony:** p. 22t, 25b, 81i, 82, 85t; **Gordon Grigg:** p. 21t, 52b, 86; **Allen Greer:** p. 7b, 49b; **PRTF:**
pp. 80-81; **Brad Maryan:** p. 14t, 19, 33b, 39, 43b, 46, 51i, 57b, 59b, 65t, 84t; **ANT/Gerard Lacz:** p.
16b; **ANT/J.Frazier:** p. 63b; **Martyn Robinson:** illustrations p. 15, 17; **National Philatelic Collection,
Australia Post:** for permission to reproduce the stamp shown on p. 29.

CONTENTS

An Introduction to Frogs

Many species of frog around the world are endangered or have become extinct.

*P*eople are becoming much more aware of the number and diversity of frogs that live in Australia today and are far more concerned about their welfare. This could be the result of increased media attention on the decline and the disappearance of so many frogs in recent years.

What Makes a Frog a Frog

Like birds, mammals, reptiles and fish, frogs have a backbone. They are therefore vertebrates. But frogs are different from other vertebrates in that they are also amphibians. That is, they spend part of their life underwater, breathing through gills, and part on land, breathing with lungs. There are three main groups, or orders, of amphibians — frogs (tailless amphibians), newts and salamanders (tailed amphibians) and caecilians, which look a bit like large earthworms. By far the largest group of amphibians are the frogs and toads, which have some 4000 species worldwide. They are also the only amphibians native to Australia.

Frog Families

There are some 200 different species of frogs in Australia belonging to five families:
74 species of Tree Frogs (family Hylidae)
115 species of Ground Frogs (family Myobatrachidae)
17 species of Narrow-mouthed Frogs (family Microhylidae)
1 True Frog (family Ranidae)
1 True Toad (family Bufonidae), the Cane Toad, which was introduced into Australia during the 1930s.

Frogs can live almost anywhere.

Where Frogs Live

Frogs are found in most areas of Australia, even in arid and alpine regions. However, they are most common in the warm, moist tropics where there is plenty of fresh water for breeding, and an outside source of warmth to maintain body temperature. Apart from some exceptions outside Australia, frogs are not found in salt water as the salt would cause them to quickly dehydrate and die.

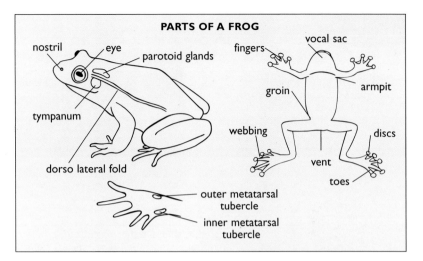

PARTS OF A FROG

nostril · eye · parotoid glands · tympanum · dorso lateral fold

fingers · vocal sac · groin · armpit · webbing · vent · discs · toes

outer metatarsal tubercle
inner metatarsal tubercle

How Frogs Grow

Frogs never stop growing, even after reaching reproductive maturity. For any group of frogs of the same species, the biggest frog in the group will also be the oldest. However, there are factors stopping a frog from becoming gigantic; indeed, it is rare to find a frog that is even slightly bigger than the maximum recorded size for that species.

There are several reasons frogs stay small. While growth in young frogs is rapid, it slows dramatically in adults. And being smaller has its advantages: less food is required to maintain body size, hidey-holes are easier to find, and being smaller also means being less conspicuous to predators.

In this book the measurements provided for each frog should be taken as the maximum attainable for that species under normal conditions.

WORDS TO KNOW

Amplexus: the sexual embrace of frogs.
Arboreal: living in trees.
Digit: a finger or toe.
Disc: the expanded, flattened end of a frog's finger or toe.
Groin: the depression formed where the hind limb joins the body.
Horny beak: the hard central jaws of a tadpole.
Larval: a stage in the development of an animal before it becomes like its parents (the tadpole is the larva of a frog).
Metamorphosis: a change in body form from larva to adult (a tadpole changing to a frog).
Parotoid: a raised gland on the skin of a frog, situated on the head behind the eye.
Permeable: a structure that allows the passage of a liquid through it.
PIT Tag: Passive Integrated Transponder Tag.
Terrestrial: living on the ground.
Tubercle: a small rounded lump on the skin.
Tympanum: a tight membrane covering the entire entrance to the ear.
Vent: the external opening through which reproductive, urinary and intestinal products pass out.
Vocal sac: an expandable sac beneath the throat which amplifies a male's call.
Webbing: fleshy skin between the fingers or toes.

What are Frogs?

All frogs have strong back legs, which are long in leapers and swimmers and shorter in burrowers.

Like all amphibians, frogs have three stages of development: egg, larva and adult. The larval stage in frogs is the tadpole.

Frogs are plumpish, with broad heads, no tails and back limbs that are longer than their front limbs. They cannot generate heat inside their bodies as mammals do but instead, like snakes, they rely on external heat sources to maintain their body temperature. They have a thin skin that both absorbs and expels water. Their skin has to stay relatively moist to allow this process to take place. Unlike birds' eggs, frogs' eggs do not have shells, but are surrounded by a protective jelly coating instead.

Not all frogs are green: the Giant Banjo Frog has yellow spots and stripes.

BREATHING IN WATER

The skin of the frog is an amazing organ. As well as breathing in oxygen through their lungs, frogs get some of their oxygen requirements out of the water that passes into their body through their skin. Carbon dioxide is expelled as the water passes back out.

What are Tadpoles?

In the life cycle of a frog, the tadpole is the larval stage — that is, the stage between egg and frog. Most tadpoles grow inside the egg for just a few days or weeks and then push their way out of the egg to swim freely in water. They then feed on the microscopic plant life in the water while their bodies grow and change until they eventually develop into frogs.

There are some interesting exceptions to this usual development. Some tadpoles develop entirely inside the egg and hatch only when they have changed into frogs. A few, such as the Hip-pocket Frog and the Gastric Brooding Frog, have very specialised tadpole development.

At the larval or tadpole stage amphibians eat mainly plants.

What Happens to Tadpoles as they Grow?

The word amphibian comes from the Greek word *amphibios*, meaning 'leading a double life'. This is an accurate description of the life cycle of a frog because the life it leads as a tadpole is completely different from the life it leads as a frog. To lead these two lifestyles requires two different types of body.

A Green Tree Frog nearing the end of its larval stage.

Some Major Changes

Tadpoles begin life as aquatic animals. Like fish, they have tails but no limbs, and they absorb oxygen through their gills from the water they swim in. As they grow, they begin to develop legs and their tails begin to be slowly absorbed into their body. At the same time they change internally. They develop lungs and, towards the end of their tadpole life, they begin coming to the surface to gulp air because their gills can no longer produce sufficient oxygen.

The tadpole's digestive system also changes to allow for a change in diet from algae to insects and other small animals. Even a tadpole's skeleton undergoes extensive modifications to enable it to change into a frog. Finally, when its limbs are fully developed and its tail is almost absorbed, the tadpole leaves the water to start its 'second life' on land — as a frog. A few days after this its tail becomes completely absorbed.

Why Don't Frogs Have Tails?

Frogs belong to a group of amphibians known as tailless amphibians, as distinct from the newts and salamanders which belong to a group known as tailed amphibians. The first amphibians to come onto the land walked rather like newts and salamanders do today and, like them, had tails. At some stage in their evolution the hind legs of some of these amphibians became longer and more adapted to leaping, their backbones shortened and became stronger, and their tails disappeared. In the process of losing their tails and becoming jumpers, frogs were able to exploit habitats that would have otherwise been unavailable to them.

Frogs and salamanders share a common ancestry.

How do Frogs Breathe?

*F*rogs are air-breathing animals that take in oxygen and get rid of carbon dioxide just as we do. Like us they have lungs which do most of this work, but they also have two other ways of breathing — through their skin and through the lining of their mouth cavity. The skin of most frogs is thin so that it can easily absorb oxygen and expel carbon dioxide, provided the skin remains moist. In the same way gases are also exchanged through the lining of the mouth. Frogs take air into their lungs by raising and lowering their throat, which draws the air into and out of their mouth through their nostrils.

Hibernating frogs breathe through their skin.

Are Frogs Slimy?

*M*any people believe that frogs are slimy little beasts, and some even believe they'll get warts if they touch them. In fact, only a few frogs are actually slimy, most are just wet and slippery to touch, rather like a fish just out of the water. A few frogs even feel almost dry.

A frog's skin is wet so that it can absorb oxygen from the air through its skin. Its skin is covered with glands, and some of these glands secrete a type of mucus that keeps the skin wet and cool. Other glands secrete a substance that is toxic or noxious to predators and another set of glands helps to keep the frog waterproof.

Not slimy — just cool, wet and waterproof.

Do Frogs Drink?

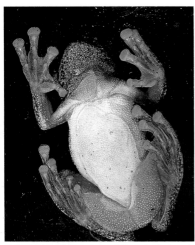

Like this Peron's Tree Frog, all frogs drink through their skin — not their mouths.

*F*rogs do not drink water, but take it in through their skin when they are in moist conditions. The problem is that they also lose water through their skin when in dry conditions. As they have to come out of the water to catch food, mate and colonise new areas, they have developed strategies to cope with this problem.

Some species absorb and lose water more rapidly through the skin on their stomach than on their back, which is most exposed and therefore more prone to evaporation. In others the skin on the back is waterproofed to help prevent water leaving the body. Some species can absorb water from the soil when the conditions are right. Once in contact with water, a dehydrated frog absorbs water in a much shorter time than it took to lose it.

Do All Frogs Swim?

*A*ll frogs can swim and some are excellent swimmers, but those that live in habitats without permanent bodies of water are usually poor swimmers. The good swimmers have more webbing on their feet than the poor swimmers.

The strong swimmers use only their hind feet, which are both moved together, to propel themselves through the water, and they keep their 'arms' against the sides of their body. The less able swimmers use a stroke like a kind of dog paddle.

Frogs may be able to swim but they are also able to drown. If a frog gets into a body of water, such as a swimming pool, and can't clamber out then it will eventually drown.

Champion swimmers like this Leaf Green Tree Frog dart across the water using only their hind legs.

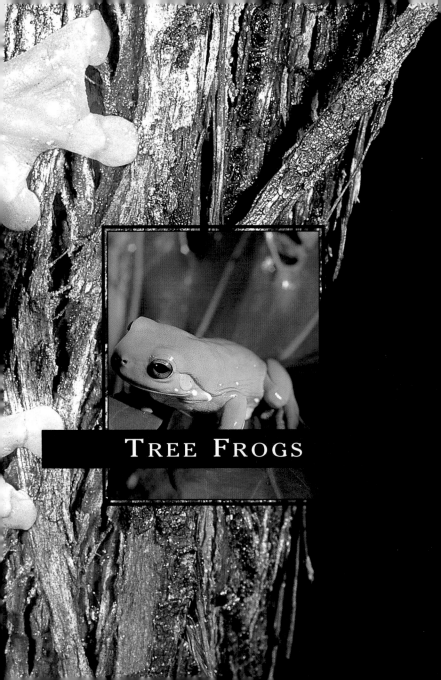

TREE FROGS

Who are the Tree Frogs?

When most people think of tree frogs, they picture a colourful frog with a flattened body, long legs and intriguing expanded discs on its fingers and toes. In fact, the 74 Australian species in this family are quite diverse in lifestyle and appearance.

Tree frogs, or hylids, are found in many countries but the greatest variety live in Australia, New Guinea and South America. They are split into three groups

Most people expect tree frogs to resemble this Peron's Tree Frog.

— the climbing frogs, the ground-living frogs and the burrowing, or water-holding, frogs. Despite their different lifestyles, these frogs have been classified by scientists as one family due to the presence of particular bones in their limbs.

Do Tree Frogs Really Live in Trees?

Some tree frogs live in trees, but certainly not all of them. Most of the climbing frogs live above the ground, but many live next to streams and ponds and never climb much higher than a metre in reeds or low shrubs.

The Green Tree Frog is adept at climbing trees.

Other tree frogs do spend a lot of time in trees and, in some cases, they will climb to quite a height above the ground. The Green Tree Frog, for instance — when it is not living in toilets and letterboxes — often lives in the hollow limbs and trunks of large gums along the banks of rivers or creeks. From these hollow hideouts the sound of its calling reverberates dramatically. Several species of the smaller, green-coloured tree frogs even take up residence in the crowns of banana plants, often concealed between the bananas themselves. As a result of this habit, the occasional frog is carted off to market in a fruit truck.

How do Tree Frogs Stick onto Trees and Rocks?

*T*he secret of the frog's ability to stay stuck to vertical surfaces lies in the expanded pads, or discs, on the end of each finger and toe. If you looked through a microscope you would be able to see that the surface of these pads is covered with inter-locking cells separated by narrow gaps.

These help the frog in two ways. If the surface they are leaping onto or climbing up is rough, the discs catch onto the irregularities on the surface. On smooth objects, they act rather like a suction cup, creating surface tension between the surface and the frog itself. The frog also presses its chest and stomach onto the surface to provide a much greater area of contact.

These 'handy' little discs act like suction cups on smooth surfaces.

How do Tree Frogs Rest?

*I*f you are a frog and you live above the ground, then you need some special adaptions so you don't fall off when you are resting. It is no good being squat and round if you want to rest, whether vertically or horizontally, on a branch. Keeping your balance would require far too much shifting about. (Imagine trying to stick a golf ball to a wall.) A longer, flatter form has far less trouble. Tree frogs have a flattened shape and are slimmer and longer than ground-living frogs. When they are at rest they take up a distinct posture, with their body compressed onto the surface and their limbs carefully tucked in under the body. This improves their adhesion to the surface and reduces water loss as less body surface is exposed to atmospheric drying.

Just as important, this posture reduces the size of the frog's silhouette so that it merges with the surface it is resting on and is less easily spotted by predators.

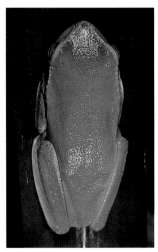

Like a fridge magnet, this Dainty Green Tree Frog won't fall off.

The Giant Frog is common in northern Australia.

Giant Frog 10 cm

The Giant Frog is one of the largest of the Australian frogs. It is found in the northern regions of Western Australia, the Northern Territory and Queensland. This frog is usually brown but some have green backs. Although this is a tree frog, it generally prefers life on the ground. Unusually for a burrowing frog, it basks in the sun at the water's edge and seems to be able to cope with high temperatures. It is normally seen only during the wet season. The tadpoles of the Giant Frog also share this tolerance for heat and have been found in water over 40°C.

The Giant Tree Frog does not seem too fussy about what it eats — any animals small enough for it to swallow, including other frogs, will do. This frog lays thousands of eggs that are fully developed in about a month. Despite the number of eggs laid, few make it to maturity because of heavy predation by other animals.

Little is known about the Short-footed Frog.

Short-footed Frog 4.5 cm

Aptly named for its short feet, this small burrowing frog is prettily marked with large dark brown patches on a yellowish background. As it is a burrower, the Short-footed Frog does not have adhesive discs on the end of its fingers and toes, and the amount of webbing between its toes is small.

Although this frog occurs all the way along the eastern coast of Queensland, very little is known about its habits because it spends most of its time in its burrow. After rain in the summer months, the Short-footed Frog usually emerges above ground and calls from the edges of ponds with a sound rather like a drawn-out growl. Tadpoles of this frog grow very quickly — they are fully developed within a month.

Main's Frogs are found in very arid areas.

Main's Frog   4.5 cm

This stout little frog inhabits some of the most arid regions of Australia, ranging through Western Australia and across the Northern Territory and South Australia. Main's Frogs have been found sheltering up to 30 cm underground. These frogs live in flat, open country that is subject to seasonal flooding, and probably only emerge above ground when the floodwater has soaked down to their underground chambers.

Main's Frogs vary in colour from green to brown, and some can actually turn from green to brown quite quickly to match their background. Feeding mainly on ants and termites, their diet extends to any small invertebrates they can catch. Their tadpoles take on a pink colouration with a pearly sheen as they develop.

The Water-holding Frog often sits partly submerged in water.

Water-holding Frog   7 cm

Even though Water-holding Frogs can burrow to a depth of 1 m, Aboriginal people were still able to find them in the desert and drink the water they held in their bladder. It has been estimated that some of these frogs can remain dormant underground without feeding or emerging for at least five years.

Their ability to hold water allows these frogs to live in arid areas, but they are just as able to live in water. They are grey and light green in colour and have a flattened head with small eyes. They often sit in pools of water with just their eyes and nostrils visible. Females lay up to 500 eggs at a time, usually in large areas of shallow flood water. The tadpoles grow more than 6 cm in length before they complete their development. This frog occurs in all mainland Australian states except Victoria.

What do Frogs Eat?

Frogs will eat anything they can fit in their mouths.

Although tadpoles are essentially vegetarian, frogs are carnivorous — usually eating insects and other invertebrates. Larger frogs have also been known to include small birds, lizards, other frogs, snakes, mice and turtles in their diet.

Most frogs will eat anything that they can stuff into their mouths, provided the food isn't dead, poisonous, hairy, bad tasting or stinging. A few, usually small to medium frogs, are specialist feeders and dine exclusively on termites or ants. Many species are opportunistic feeders, and what they eat depends on what is available at the time. So if there is an abundance of crickets they might eat those exclusively while they are available and then switch to something else when the supply runs out. Most frogs consume their prey on land and only a few catch prey while they are in the water. Most frogs have good eyesight with which to spot a meal, although normally prey has to be moving to be recognised as a food item.

Do all Frogs Have Tongues?

Only a few species do not have tongues, and all Australian frogs have them. Frogs' tongues may be very long, almost as long as the frog in some cases, and covered in a sticky substance. Unlike a human tongue, which is attached to the rear of the mouth, a frog's tongue is attached to the front, and is folded on the bottom of the mouth with the tip pointing back to the throat. From this position it can flick out very quickly and accurately to catch passing prey. The frog then draws its tongue back into its mouth, complete with food attached. Frogs' tongues are also covered with many taste buds so that when the frog doesn't like the taste of something it can spit it out.

A frog's tongue is attached to the front of its mouth.

How do Frogs Swallow?

Eye sockets help a frog swallow its food.

*F*rogs swallow their food whole. This is why they normally only catch prey that will fit into their mouths. However, sometimes a frog will see only the end of something and so catch it and try to eat it. Eventually the frog has to spit out the over-sized morsel.

Believe it or not, the frog's eye sockets help it to swallow its food. A frog can compress its eye sockets inwards and downwards against the roof of its mouth and this helps push the food towards the throat. Some of the larger frogs also use their hands to push bigger prey into the mouth, especially when they catch some insect that is all wings and limbs. Using their hands helps get all these awkward bits into the mouth.

Do Frogs Have Teeth?

*S*ince frogs swallow their food whole, teeth aren't very important. Some frogs are entirely toothless, and others have teeth only on the upper jaw. Scientists think that the only reason frogs have teeth at all is to help hold the prey until it is well and truly in the mouth. They might also be used to crush and perforate the victim. While humans have different types of teeth dedicated to chopping and grinding, frogs have no need for this and their teeth are all small conical pro-tuberances. Besides any teeth they may have in the jaw, many frogs have some additional teeth on the roof of the mouth near the front of the upper jaw. Again these seem to be simply an aid for holding the prey before it is swallowed.

> **FINE DINING UNDERWATER**
> Very few frogs catch their prey under the water, where a sticky tongue would be no use at all for snaring prey. Those few that do, simply open their mouths and grab it. They then push the food item into their mouth with their hands.

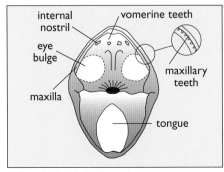

A dentist's view of a frog's mouth.

Green and Golden Bell Frog

8 cm

The acrobatic Green and Golden Bell Frog has no trouble balancing on a thin stem (above), or even a vertical bulrush (left).

Many people knew little about the endangered Green and Golden Bell Frog until a campaign was launched to save a large population in the State Brick Pit at Homebush Bay, the site for the Sydney 2000 Olympics. Upon discovery, engineers modified their plans for the site to preserve this important habitat.

Habitat

This frog prefers to live in permanent ponds containing bulrushes. Originally its range was from Victoria up the coast to northern New South Wales, with populations in the Southern Highlands and west of the Great Dividing Range in Bathurst and Orange, but it is now believed to be extinct in the Highlands, and west of the Great Dividing Range. Only small populations remain along the coast, mostly south of Sydney.

DISAPPEARING FAST

Once so common that it was collected as food for captive snakes and for class dissections at some universities, the Green and Golden Bell Frog has suffered a catastrophic decline. There seem to be several causes for this: the introduction of Mosquito Fish that eat the tadpoles, the reclaiming of swamps where these frogs live, and a fungal disease.

Diet and Behaviour

The Green and Golden Bell Frog is active by day and often basks in the sun. Feeding mainly on cockroaches, crickets and grasshoppers, it also preys on other frogs which it finds by their calls.

Females lay 2000–4000 eggs and tadpoles can take from 2–12 months to fully develop into frogs.

Green Tree Frog

11 cm

Green Tree Frogs have a close association with humans.

Everyone recognises the Green Tree Frog which has the benign, placid look of a little green buddha. It is one of the biggest of the tree frogs.

Green Tree Frogs often inhabit mailboxes, meter boxes, bathrooms and toilets. They have a particular liking for toilet bowls, making a sudden appearance when the toilet is flushed. In their natural environment they hide in hollow tree limbs and in rock crevices. They are popular pets and, given the right conditions, thrive in captivity.

Diet

A Green Tree Frog eats mainly insects, but also small birds, bats and mice. It will take anything that fits into its gaping mouth, often shoving it in with its hands. Around the home, it soon learns to take advantage of any outside lights to catch the insects they attract.

Declining Populations

Unfortunately, populations of the Green Tree Frog have declined noticeably in some areas and have all but disappeared from the Sydney region where the frog was once common. This is probably due to the removal of old water tanks and the filling in of ditches and temporary ponds.

> **A SLIMY CURE**
> The Green Tree Frog has large glands behind its head which contain a slimy secretion. When this secretion was analysed it was found that one of the components, caerulein, could lower blood pressure in humans. Caerulein is now manufactured artificially and is used in several clinical applications.

19

Cave-dwelling Frog   5.5 cm

This frog is a relatively recent find.

As its name suggests, this frog lives in caves, although it is also known to take refuge in the gaps between large boulders. In regions where the Cave-dwelling Frog is found, temperatures can be high — often exceeding 40°C — so it is not surprising that this frog shelters in the cooler, more humid environment of caves.

Although this frog bears some resemblance to the Green Tree Frog, it is about half its size and not as heavily built. Its skin is covered in small lumps that give it a grainy texture.

Scientists have only known about the existence of the Cave-dwelling Frog since 1979, when it was found on the Mitchell Plateau in the Kimberley region of Western Australia. Nothing is yet known about the frog's habits, how it breeds or how its tadpoles develop.

Magnificent Tree Frog   10.5 cm

A huge gland covers this frog's head.

It is quite surprising that the Magnificent Tree Frog was not discovered until the 1970s, since it is one of the largest of the Australian frogs and certainly one of the most impressive.

This frog has a huge gland that covers the whole of the top of its head, which makes it look like its brains are bulging through. These frogs are not often encountered as they occur mainly in the north-western area of Western Australia and adjacent regions in the Northern Territory. Of those that have been sighted, most were found in public toilets. Apart from the fact that they hide in caves and rock crevices, little is known about their habits. Females probably lay several thousands of eggs and deposit them in several different pools of water to increase the chances of at least some of the tadpoles developing into frogs before all the pools dry up.

Tasmanian Tree Frog
6 cm

Tasmanian Tree Frogs live only in Tasmania. They can be found throughout the western region where they inhabit the vegetation growing around pools. These frogs are also found many metres off the ground in eucalypt trees, so they are obviously very accomplished climbers.

These frogs breed during spring and summer, and clusters of eggs are attached to any vegetation that is growing in the water. Most of their tadpoles complete their development

The Tasmanian Tree Frog is inactive during winter.

and emerge as frogs in summer and autumn. In some cases, however, tadpoles delay their metamorphosis until the following spring to take advantage of the warmer weather and the availability of more food. The call of the male frogs sounds similar to the honking of geese.

Dainty Green Tree Frog
4.5 cm

This pretty little frog from coastal Queensland and northern New South Wales is often called the Banana Frog because of its habit of hiding among bananas and ending up in fruit shops or homes in southern Australia. It is difficult to see on vegetation because it presses itself very close to a surface and tucks its limbs tightly against its body so its fingers and toes are hidden. In this way the yellow colouration on the side of its body and limbs is also hidden. As well as being a great camouflage, this pose also helps the frog to avoid dehydration because less skin is exposed.

The Dainty Green Tree Frog is usually found in dense vegetation at the edge of water, and in ditches and swamps. It breeds in shallow water such as flooded grassy areas and small ponds where the eggs are attached to grass stems and similar plants.

A Dainty Green Tree Frog in typical repose.

How do Frogs Mate?

*M*ales call, often in chorus, at breeding sites in the hope of attracting a female. Females respond only to the calls of males of their own species and pay no attention at all to the calls of others. They don't just choose the first male of their species they hear, but spend time listening to several before they make their choice.

Once the female approaches, the male moves behind her and grips her around the body with his arms, a position known as amplexus. Depending on the species, he holds her either above the arms or around the waist. In many species the male has spines or pads on his hands that probably improve his grip on his mate.

A male grips a female in amplexus.

How are the Eggs Fertilised?

*E*ggs are fertilised outside the female's body. The female deposits her eggs, each surrounded with a coating of jelly, and at the same time the male expels sperm onto the eggs in order to fertilise them. In some cases the male or female holds the mass of eggs with their legs as they leave the female, probably to make sure that the eggs are fertilised.

Eggs may be laid in still water, flowing streams, in moist places on the ground, or underground, depending on the species. They are laid in long strings or singly in a frothy mass. The number of eggs laid depends on the particular species and can range from just five or six eggs to many thousands.

A female expels her frothy mass of eggs.

NATURAL SUN SCREEN

Some frog eggs have a black pigment on their upper side that screens out the sun's ultraviolet rays which could otherwise harm the embryo. This pigment occurs only on the eggs of species that deposit their eggs in open areas where they are likely to be exposed to sunlight.

When do Frogs Breed?

*I*n temperate regions where there is permanent water, or at least consistent rainfall, many frogs breed almost all year round. In harsher, more arid regions they breed only at specific times.

Burrowing frogs only breed after heavy rain.

Frogs that breed only in certain conditions are sometimes called explosive breeders, because their urge to breed usually needs some kind of trigger. For example, breeding in burrowing frogs is triggered by water soaking through the soil down to the level of their burrows. This indicates that there is enough surface water for the tadpoles to survive until they metamorphose. Although water is the critical factor for most seasonal breeders, many arid-zone frogs ignore winter rains and breed only during summer rains.

Are Frog Eggs the Same as Other Eggs?

*U*nlike the eggs of birds or reptiles, frogs' eggs don't have a shell or any other type of membrane. As they are laid, each egg is coated with a mucus-like substance which swells on contact with water and forms a protective jelly around the egg. The eggs contain a yolk that provides nourishment while the embryo develops. Waste products from the embryo pass through the jelly into the water.

Birds and reptiles can lay their eggs in dry places because the protective shell reduces water loss, but frogs' eggs have to

Eggs are protected by a jelly-like coating.

be laid in water or other very moist places. Because the protective jelly does not last long in water, embryos usually develop into tadpoles very quickly.

Green-thighed Frog 4.5 cm

These frogs are found only after heavy rain.

This handsome frog has a deep chocolate brown back with bright green or blue–green and black on the groin and the back of the thighs, and a black stripe from the snout to shoulders.

Because this frog is hard to detect, its habits remains something of a mystery. Most sightings of this frog have been made after heavy rain during the warmer months when the frogs are involved in breeding activities. At these times they are sometimes seen in flooded paddocks or temporary ponds and ditches or in the vicinity of wet forests or woodlands. The Green-thighed Frog is regarded as a vulnerable species in New South Wales, and as a rare species in Queensland, although it may just be very hard to find unless conditions are right.

Broad-palmed Frog 4 cm

The Broad-palmed Frog lays eggs almost anywhere.

This frog looks quite similar to several other tree frogs. It is usually grey or light brown. It has a dark, wide band running from the nostrils to the eyes, where it is broken by a white bar before continuing from the eyes to the shoulders. These agile, ground-dwelling tree frogs, found in northern New South Wales, Queensland and South Australia, live in all types of woodland and open areas. They shelter under rocks and wood or in deep leaf litter, and are often found quite a distance from permanent water. During the breeding season they move close to any sort of water, whether natural or artificial, flowing or still. Breeding begins in spring after rain and continues through summer. Females deposit their eggs in clumps among weeds and grasses in shallow water.

Desert Tree Frog

4 cm

This small frog ranges from light grey to a red–brown with a dark band along the side of its head and body. It is one of Australia's most widely distributed tree frogs, and it also occurs in New Guinea.

A male calls while a pair mates.

Desert Tree Frogs congregate, often in very high numbers, around any permanent water. In the gorges at Mutawintji National Park in western New South Wales the noise of Desert Tree Frogs calling at night is absolutely deafening. At times they are so thick on the ground that it is almost impossible to walk around the pools without treading on them.

These frogs occur not only in arid areas but also in moist coastal regions and they often live close to humans, taking up residence, alongside the Green Tree Frog, in buildings, swimming pools and toilets. Females lay up to 300 eggs that develop into frogs in 1–2 months. Tadpoles are very tolerant of high water temperatures, which is just as well because the water can be as high as 40°C.

Waterfall Frog

5.5 cm

The Waterfall Frog inhabits parts of northern Queensland and is a mottled green–grey in colour with black patches.

This fascinating frog has an array of remarkable adaptations for surviving amongst the slippery rocks in cascading waterfalls and rushing streams. Because a complicated mating call is hard to hear when surrounded by the noise of fast-flowing water, males have dispensed with a vocal sac

This frog's call sounds like a growl.

and their call has been reduced to a growl. Males have small spines on their thumbs and chest which they use during mating to cling onto the back of a female so that they won't be dislodged by the force of the water.

Females lay large eggs under stones in the stream, surrounding them with a sticky mass to keep them in place. Tadpoles have a sucker mouth that allows them to cling to slippery rocks and so avoid being swept downstream.

How Long Does it Take From Egg to Frog?

A frog starts off as an egg which hatches into a tadpole which then metamorphoses into a frog. The length of time this takes varies considerably between species and depending on conditions. Some tadpoles emerge from their eggs in a few days, others take anywhere from a few weeks to several months.

Several factors influence tadpole development.

The time it takes for a tadpole to become a frog is just as variable. It may take just two weeks or as long as 12 months. Tadpoles in arid regions have to race to develop before the water dries up and many of them don't make it. Other species remain as tadpoles over winter and complete their development the following spring when insects are more plentiful.

Do Frogs Care for their Eggs?

F or most frogs, parental responsibilities end when the eggs have been laid, and tadpoles are usually on their own. But there are some exceptions. Several frogs lay only a few eggs at a time on land and in many of these cases an adult stays on or next to the eggs until the tadpoles or young frogs emerge. In some species this is the male's job and in others it is the female's.

As frogs have glands in their skin that contain antibiotics, it may be that the presence of an adult is essential to protect the eggs from fungal infections so they hatch successfully. The adult may also be there to guard the eggs from predators or to keep them moist.

REAL DEVOTION

Some species of frogs outside Australia exhibit extraordinary egg care. A South American frog carries her eggs around in a pouch on her back. A Jamaican tree frog lays her fertilised eggs in water in the crown of a bromeliad, then returns and lays unfertilised eggs for the tadpoles to feed on. No Australian frogs are known to behave like this.

The Red-crowned Toadlet guards its eggs.

Do Frogs Care for their Tadpoles?

*P*arental care of tadpoles is even less common than parental care of the eggs. Only three species in Australia are known to play a part in the development of their young.

Two of these, the Northern and the Southern Gastric Brooding Frogs, carry the developing tadpoles in their stomach and release them through their mouth as frogs. The male Hip-pocket Frog carries the tadpoles around in pouches on its side. The tadpoles emerge from the eggs and then clamber and wriggle over the adult until they get into the pouches. There they stay until metamorphosis is complete, when they emerge and disperse.

Taddies with daddies: the remarkable Hip-pocket Frog.

Do Frogs Build Nests?

*S*ome Australian frogs lay eggs in existing burrows or construct new burrows or shallow chambers especially for the eggs. The Sandhill Frog and the Turtle Frog both lay their eggs deep underground. They probably construct some sort of chamber for them to develop in, but what form this takes is unknown.

Other frogs create nests of foam out of air bubbles and mucus to contain their eggs, and place them on the ground under litter, in moss, or in a burrow. The purpose of the foam may be to maintain moist conditions during development. Some species make their nests in water where they float on the surface. Their purpose is probably to speed up the development of the eggs because at surface level the water temperature is warmer and oxygen more abundant.

Some frogs create foamy nests to hold their eggs.

Eastern Dwarf Tree Frog

2–3 cm

In slippery vegetation the little Eastern Dwarf Tree Frog is agile and sure-footed (above). A male calls for a mate (right).

This slender frog is one of the smallest of the green tree frogs. It ranges in colour from green to fawn, and has orange to yellow patches on the groin and the back of the thighs. It also has a distinctive white stripe running from its jaw down to its front limbs. Its eyes are golden.

Habits and Habitat

The Eastern Dwarf Tree Frog lives in coastal areas in vegetation along the edges of swamps, ponds and streams, sometimes in large numbers. Males call from reeds or other vegetation in the water both day and night during the warmer months. Females lay about 250 eggs in a number of clumps attached to vegetation in still water. Development into frogs takes about 100 to 120 days.

Like the males of several other species, male Eastern Dwarf Tree Frogs practise a series of arm and leg waving movements that scientists call 'foot flagging'. At first it was thought foot flagging was used by males to attract females. However, recent studies have found that flagging occurs between males and this habit is now thought to be a part of male–male aggression practices. Apparently, only the loser makes the movement, so it may be an appeasement gesture.

Blue Mountains Tree Frog

6 cm

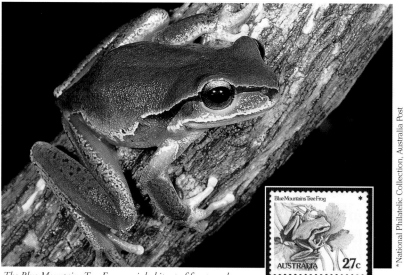

*National Philatelic Collection, Australia Post

Blue Mountains Tree Frog *

AUSTRALIA 27c

The Blue Mountains Tree Frog, an inhabitant of forests and streams (above) and part-time portrait sitter (right).

This is indeed a handsome frog. Its back and the top side of its legs vary in colour from green to brown and its sides bear patches of green. It has a black-and-white stripe running along both sides from its snout to its groin. Its groin and the back of its thighs are bright red to orange in colour.

> **STAMP OF APPROVAL**
> This celebrity frog appears on one of a series of seven Australian postage stamps featuring amphibians and reptiles. The other frogs featured on the stamps are the Corroboree Frog and the Crucifix Toad.

Distribution and Breeding

As the name suggests, the Blue Mountains Tree Frog occurs in the Blue Mountains, west of Sydney, and across a much wider area south to Victoria. Because it inhabits forested areas and places close to flowing streams — particularly where there are large boulders — it is not often seen. A strong jumper, it hides in crevices or under rocks during the day.

The males call during spring and summer from vegetation and other positions close to the water. The call is a sharp scream followed by a soft trill. The eggs, which can number up to 1000, are deposited on rocks in the flowing water. Tadpoles take 2–4 months to complete their development.

29

BURROWING GROUND
FROGS

Who are the Ground Frogs?

Ground frogs are found only in Australia.

*G*round frogs, also known as southern frogs, are a family of frogs that exist only in Australia. They are related to the Leptodactylidae family of South and Central America. The ancestors of both families once lived on the great southern continent of Gondwana.

Although they all live on the ground, they are still a very mixed bunch. They are found right across Australia and live in many different types of habitat. Many live in arid areas, some prefer alpine regions, and others inhabit moister areas. The type of aquatic environment they occupy varies from swamps, seepages, roadside drains and still water ponds to rushing torrents. Some species are completely aquatic. A few show parental care of eggs or young, including two (the Northern and Southern Gastric Brooding Frogs) whose young develop in the stomach.

Where do Ground Frogs Lay Their Eggs?

ALL IN A LATHER

Some female ground frogs, when laying their eggs, beat on the water with their front feet to create air bubbles which are then pushed under the body. The bubbles are trapped in the mucus which the female discharges along with the eggs, creating a frothy mass. Many females have toe fringes that help in this frothing and beating process.

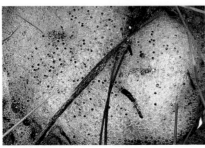

The black dots in this froth are eggs.

*D*ifferent species of ground frogs have different ways of ensuring that their young have the best chance of developing into adults. Some species coat their eggs with a protective jelly, while others lay them in a mass of foam, a bit like whipped egg-white. Some ground frogs lay their eggs in water where the tadpoles emerge and swim freely, others lay them on land and the tadpoles emerge when the eggs are washed into water. The tadpoles of some species have to rely entirely on their egg yolk for food and develop completely within the egg, emerging only as fully developed frogs.

One species carries around its developing tadpoles in pockets on its body, and another swallows its eggs so the tadpoles can develop safely within its stomach.

Are There Any Frogs in the Desert?

*M*ost people associate frogs with permanent water supplies and so they are surprised to learn that some frogs have adapted to a dry environment and can thrive in the desert regions of Australia.

Desert-dwelling frogs have developed various strategies for coping with their harsh environment. Some remain near permanent waterholes and springs that provide them with a suitable micro-environment. Others, however, stay deep beneath the dry sandy beds of temporary watercourses, or disperse themselves across the floodplains, burrowing into depressions such as claypans. Most of these burrowers belong to the ground frog family but one group of tree frogs has adapted to desert living by adopting the same practice.

Burrowing ground frogs emerge after rain to feed and breed.

How do Desert Frogs Survive Without Water?

*R*eal desert frogs have a range of adaptions for survival in the desert when conditions get tough. All species are burrowers and they use the soil to insulate themselves from heat and to avoid dehydration.

Water-holding Frogs absorb large quantities of water through their skin and store it in their tissues. Other species retain water in their large bladders which are capable of storing water weighing up to half the frog's body weight. Desert frogs are quick to breed when there is enough rain to form temporary bodies of water. Almost as soon as the rain stops they begin calling and mating. Tadpoles develop rapidly and can survive high water temperatures.

A Water-holding Frog fit to burst.

QUICK OR DEAD
Tadpoles of desert-dwelling frogs have to rapidly develop into frogs so that conditions are still favourable. If they don't, they risk dehydration and death when the pools of water dry out.

Trilling Frog

5.5 cm

A Trilling Frog burrows backwards into the sand (above), and submerges itself within minutes (left).

One of the most widely distributed of the Australian burrowing frogs, the Trilling Frog inhabits grasslands in arid and semi-arid regions, often in large numbers. The Trilling Frog is so good at burrowing that, on loose soil, it can completely disappear beneath the surface in a few minutes. This frog is often found in claypans that fill with water after heavy rains.

Fawn to grey with darker patterning, individuals often have a paler stripe down the middle of their back. Trilling frogs eat large quantities of ants and termites but they also eat spiders and beetles when they get the chance.

Dangers

Males call, usually after heavy summer rain, while they float in the water, and females lay their eggs in the shallows. Tadpoles develop quickly before the pools dry out. Some don't make it in time, and dessication is the most common cause of death. The tadpoles seem to be safe from waterbirds, perhaps because they taste unpleasant or are poisonous. Small, newly developed frogs, however, are preyed upon by beetles, ants, snakes and foxes. Adults are vulnerable when they are on the surface after rain and are eaten mainly by foxes, cats and snakes. Because they are active only at night, they are rarely taken by birds. Researchers think they feed only after heavy rain when they emerge from the ground, the time when insects and other invertebrates are active.

> **HOW TO LIVE A LONG LIFE**
> Longevity is vital to arid-zone frogs. They depend on warmth and water to breed, conditions that only arise every few years. They need to live long enough to breed in the next spell of rain.

Crucifix Toad

6.5 cm

The colourful markings on this Crucifix Toad (above) resemble the shape of a crucifix (right).

The Crucifix Toad is also known as the Holy Cross Toad as it has what looks like a jewelled cross growing on its yellow back. The 'jewels' are actually small black, red, yellow and white bumps. This frog also has glands on its back that release a sticky fluid which is possibly toxic, or at the very least tastes unpleasant to potential predators. It has short limbs, a small head and a stout body which make it look rather like a slightly squashed golf ball. Despite its build, the Crucifix Toad can move around quite rapidly.

Habitat and Diet

This frog is normally found in grasslands and seems to prefer black-soil habitats. Despite the harsh environment, it occurs in large numbers in the right conditions — usually after prolonged late summer rain. It feeds almost entirely on ants and termites which are plentiful after rain.

Behaviour and Reproduction

This is one of the species of frog that burrows vertically into the soil in a circular fashion, rather like a corkscrew. These frogs spend most of their life underground, coming to the surface to eat and breed only after heavy rain. The male calls as it floats spread-eagled in a pool of water and sounds a bit like an owl. Tadpoles develop quickly so that they can take advantage of the temporary abundance of water.

35

How do Frogs Burrow?

Characteristically, burrowing frogs are stout with short arms and legs. After all, there is little point in having long jumping legs if a good part of your life is spent underground. While the tubercles or lumps under the feet of most other frogs are flattened, many ground frogs have developed tubercles with a shovel-like cutting edge which they use to scrape out soil.

Giant Burrowing Frogs are efficient and powerful burrowers.

Ground frogs burrow well below the surface where they construct a chamber. They may remain inactive in these chambers for several years, waiting for signs of sufficient rain.

To be able to breed successfully these frogs need enough water to last until their tadpoles have grown into frogs. Building burrows deep in the ground, these frogs know that conditions are right when the water seeps into their burrow. If they didn't dig so far down, they could be fooled into surfacing before there was sufficient water for their needs.

> **DIG THIS**
> Frogs do their burrowing in three distinct ways. Some frogs dive into the soil head-first at an angle, some slide in backwards at an angle, and others go in backwards digging vertically with a corkscrew action.

Why Don't Frogs Dehydrate Underground?

After bloating themselves up with water and food, burrowing down below the soil and constructing a comfortable chamber, these frogs still face a problem of sufficient moisture in arid areas. Because their skin can lose water

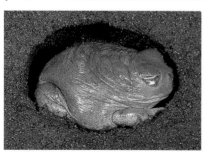

The skin cocoon prevents dehydration.

as well as absorb it, they must avoid losing their stored water into the surrounding drier soil. They do this by shedding their skin and cocooning themselves in it. The outer layer of skin separates from the body in one piece and then encloses the frog. The longer the frog stays underground the more layers it grows and sheds, each one sticking to the previously shed skins. The cocoon slowly hardens, seals in the frog and significantly reduces water loss.

How do Tadpoles Feed?

Some tadpoles feed entirely on the yolk of the egg they hatch from, but most feed on algae and other vegetable matter. Such tadpoles have a horny beak surrounded by several rows of tiny teeth. Tadpoles use these teeth to scrape off small pieces of food while larger pieces are cut up with their beak. These tadpoles also collect micro-organisms as they suck in water and filter it through their gills. Tadpoles have strands of mucus which run from their gills to their gullet. Any small creatures swimming in the water are trapped in this mucus.

Tadpoles use their tiny teeth and beak to break up food.

Most of a tadpole's body cavity is filled with its long, coiled intestines which have a section modified for digestion instead of a true stomach. Very few tadpoles are predatory, but the Sandpaper Frog is cannibalistic at a certain size — it grows faster that way.

Can Frogs Regulate Their Temperature?

A frog's body temperature usually matches that of its environment as, like reptiles and fish, it has no way of generating or retaining body heat. Despite this handicap, frogs occur in both very hot areas and areas where snow covers the ground for some months of the year. The tadpoles of some arid-zone species are able to survive very high water temperatures and, amazingly, one species has been found in water at almost 40°C. Some frogs studied overseas have been found surviving in temperatures as low as –6°C. No similar studies have been made in Australia and nothing is known about where or how Australian alpine frogs survive over winter.

Frog on ice.

Some species raise their temperature by basking in the sun, but they are usually species that live near permanent water. At least they can jump in to cool off!

The Painted Frog's call is like a musical trill.

Painted Frog 6 cm

This stout little frog inhabits regions of grassland and open woodland in South Australia, Western Australia, Victoria and south-western New South Wales. It has a blade-shaped growth on the underside of its hind feet which it uses to burrow into the ground. The sides, back and limbs of the Painted Frog are cream to yellow with dark brown or olive blotches. These frogs usually have a pale stripe down the middle of their back. Their stomachs are white.

Male Painted Frogs have backs that are covered with small rounded bumps and during the mating season these bumps develop spines, presumably to ward off any other male from embracing them by mistake. The Painted Frog, like several other species, is usually seen only in the summer months after rain, when it becomes active. The breeding season of the Painted Frog spans autumn and winter, when females lay up to 1000 eggs in a long chain among vegetation submerged in the water.

Sudell's Frog has large, prominent eyes.

Sudell's Frog 5 cm

Widespread in south-eastern Australia, Sudell's Frog inhabits open grassland and areas of woodland. This frog remains buried until there is sufficient rainfall to bring it to the surface, after which it can appear in large numbers.

Sudell's Frog is a small stout frog with large prominent eyes and short arms and legs. The body may be tan, yellow or brown with a pattern of darker blotches. There is usually a pale stripe down the middle of its back. One of its distinguishing features is the membrane of skin in the groin which extends between the knees and the side of the body. In common with many other burrowing frogs, Sudell's Frog has sharp burrowing tubercles on the underside of its feet.

After rain, males call while floating in shallow bodies of water in flooded areas. During this time they develop small black spines on their back, possibly to deter other males. Females lay chains of up to 1000 eggs in shallow flood waters.

This frog shelters more than a metre underground.

Desert Spadefoot Toad   6.5 cm

The Desert Spadefoot is closely related to the Crucifix Toad, although this species lives in central and western parts of Australia and is not in contact with its cousin. It has warty skin and is dark grey to brown in colour with scattered white, yellow, red and black spots. It is named after the spade-shaped tubercle on its back feet, which it uses to help it burrow. Individuals have been found more than a metre underground.

The Desert Spadefoot Toad is often active on humid nights when it feasts on ants and termites which are plentiful in its habitat. (One frog was observed to eat more than 300 termites in one sitting.) The glands on its back ooze a noxious, sticky, whitish fluid that probably tastes unpleasant and so deters predators.

The female Desert Spadefoot Toad lays her chains of eggs in shallow, temporary pools where they become entwined with submerged grasses. Like other desert species, their tadpoles mature quickly, and are mature in about 16 days.

Shoemaker Frog   5 cm

This plump little frog gets its name from its call, which resembles a series of short taps something like the sound of a shoemaker tapping nails into a shoe. A beautifully patterned frog, it is gold or bronze in colour splashed with dark brown blotches. Its legs are relatively short when compared to the size of its body, and its skin is smooth and moist. Breeding males do not develop spines or lumps on their back like many of their close relatives.

The Shoemaker Frog occurs in Western Australia, spreading into the border regions of South Australia and the Northern Territory. It lives

The Shoemaker Frog's name comes from its call.

in areas of grassland, where it usually breeds in water-filled claypans. After a spell of heavy rain during the summer months, males call to attract females for mating. Females lay their eggs in a long string in the water. Tadpoles take almost six weeks to complete their development into frogs.

Pobblebonk

8 cm

Pobblebonks can be variable in their skin colour and patterning, and may represent several subspecies (above and left).

The call of this frog sounds like a single 'plonk', rather like a string being plucked on a banjo. When one frog starts to call, others join in and rapidly repeat the single note over and over again. Because each frog calls at a slightly different pitch, when they call together they sound like a chorus.

Pobblebonks are active during and after rain and are usually found near still or very slow-moving permanent water. Male Pobblebonks call throughout most of the year from vegetation in the water, and sometimes congregate in large numbers. Female Pobblebonks often lay their eggs in a foam nest, frequently laying more than 3500 at a time. Tadpoles usually take 5–6 months to develop, although in certain conditions they can take longer.

HIGH LIVING

In the Snowy Mountains Pobblebonks have been found as high as 1600 m above sea level. At such altitude, the tadpoles take 12 months to metamorphose and frogs can grow to over 8 cm in length. Tasmanian Pobblebonks have never been found at high altitudes despite their presence elsewhere.

Description

Pobblebonks can vary widely in appearance, being anything from brown to grey, flecked or blotched with darker colours. These frogs have white stomachs which are sometimes mottled with grey, and a pale yellow stripe starting below the eye and ending above the arm. Some have a thin pale stripe running down the middle of their back. Because Pobblebonks vary so much in colour, size and males' mating calls, these frogs have been grouped by scientists into five races or subspecies.

Giant Burrowing Frog

9 cm

Unlike the Cane Toad, this frog has yellow spots on its flanks (above), and a friendly smile (right).

The Giant Burrowing Frog is also known as the Owl Frog because its soft continuous call sounds like the call of an owl. This frog lives on the east coast of New South Wales and Victoria. It ventures out of its hiding place on warm wet nights but it is rarely seen these days as its numbers have declined.

Description

The Giant Burrowing Frog is dark grey or brown in colour with scattered yellow spots on its sides and a yellow stripe running from under each eye to its ear opening. It has coarse, warty skin and is sometimes mistaken for the introduced Cane Toad. However, this frog's pupils are vertical while the Cane Toad's are horizontal. Nor does it have the large glands behind the head which are so noticeable in the Cane Toad. Breeding males also have quite obvious black spines on the back of their fingers — most likely used to fight other males.

Behaviour and Reproduction

The Giant Burrowing Frog is an efficient and powerful burrower. Males construct burrows in the banks of creeks, although they are also known to occupy Yabby burrows. They call from their burrows, usually on wet nights, during spring and summer. The female lays her eggs at the end of the burrow in a chamber filled with water. The tadpoles are flushed into the creek when rain waters flood the burrow. Tadpoles are large and dark — up to 7 cm — and are sometimes found in deep pools along creeks containing clean water.

41

Ornate Burrowing Frog 5 cm

This frog lives in both arid and coastal regions.

This frog comes in a range of colours and patterns. Its back can be anything from pale to dark grey or brown, with or without a darker patterning. Its limbs are usually barred or spotted with darker markings and there is often a prominent light patch on its head behind its eyes.

The Ornate Burrowing Frog is found across northern areas of Western Australia, the Northern Territory and Queensland, extending down the east coast to the central west of New South Wales. It occurs in coastal areas as well as more arid regions and is an efficient burrower. It is one of the group of ground frogs that burrow in backwards at an angle. It is usually only seen when it is foraging — normally on wet nights during the spring and summer months. The female may lay more than 1500 eggs, which she deposits in water in several foam nests.

Spencer's Burrowing Frog 4.5 cm

Spencer's Burrowing Frog is a desert dweller.

Not much is known about this medium-sized burrowing frog which occurs in the desert areas of central Western Australia extending into the Northern Territory and South Australia. Similar to the Ornate Burrowing Frog, which occurs further north, it can be distinguished by the webbing that grows about halfway up its toes.

It lives in sandy areas in temporary watercourses or in other places where flooding occurs during heavy rain. Active at night, this frog burrows into the sand during the day or seeks shelter beneath timber or other debris. Spencer's Burrowing Frog has also been found breeding in rock pools in the hills, so it is obviously prepared to travel during times of good rain to take advantage of better breeding spots. The male calls from water at night. The female will lay up to 1000 eggs in a foam nest.

Giant Banjo Frog

9 cm

This big, colourful frog has a rather drab brown back scattered with darker spots but its sides are orange and black and its stomach is bright yellow. There are raised yellow to orange bands running along its sides from below each eye to its arms.

Occurring in dry sandy regions of western New South Wales, this frog is an efficient burrower and has even been found inside Mallee Fowl

The male Giant Banjo Frog in a musical mood.

mounds. This is one species that seems to have made good use of stock dams as it is often found around their edges.

The breeding season extends through spring and summer when the males call from burrows or vegetation at the edge of water. The females deposit eggs within a frothy mass inside burrows. The call is a deep single note — something like a string being plucked on a banjo.

Tanami Toadlet

3 cm

The Tanami Toadlet only came to the attention of scientists as recently as 1981. It was discovered in the Tanami Desert in the Northern Territory and adjacent Western Australia. This frog lives in red sandy soil vegetated with spinifex and mulga. It burrows into the soil and emerges after heavy rain to breed.

This attractive little frog is brown on the back and is patterned with scattered spots of red and cream.

This frog lives only in and around the Tanami Desert.

The glands behind its head are pale yellow. The Tanami Toadlet has distinctive, widely-spaced nostrils.

Because this frog lives in a remote area and is so hard to study, little is known of its life history, its behaviour, its reproduction, or even what the male's mating call sounds like.

Do all Calls Mean the Same?

*F*rogs call in order to communicate with other frogs. The calling we are most familiar with comes from the male frog during the mating season as he advertises his whereabouts to nearby females. Each species has its own particular call pattern and females respond only to the call of their

The frog's vocal sac amplifies its call.

own species. Males use a different call to warn off other males who get too close to their territory.

Both males and females have a releasing call. An unreceptive female, or another male grabbed by mistake, will give a release call if it is grasped by a male as though for mating and this message is usually enough to make the offending male release his grip.

In some species both sexes give a distress call when caught, or when in danger of being caught, by a predator. It often sounds like a scream and is no doubt intended to startle the predator into letting go.

> ### WHAT'S IN A CALL?
> Frogs make sounds by forcing air from their lungs to pass over their vocal cords. This sound is then amplified by the vocal sac, which is a pouch of skin beneath the mouth with an opening into the mouth. When calling, a frog keeps its mouth closed and forces the air backwards and forwards between the lungs and the vocal sac.

How do you Describe a Frog's Call?

*T*he call of each species of frog sounds different — at least it does to the frogs themselves. Some scream, others wail, moan, trill, tap, grunt or squelch. Calls last anything from a fraction of a second to several seconds.

As well as being the way frogs tell their own species from others and identify suitable mating partners, these calls also help researchers to distinguish various species.

Frog researchers will usually hear a frog before they see one, and so learn to recognise the different species first and foremost by ear. Some researchers create recordings of frog calls using an oscillograph, which records sound waves and reproduces them in a graphic form, called an oscillogram. Researchers can read the frequency and strength of the call off the graph, and thereby know what species of frogs are calling at a particular site.

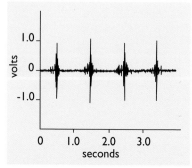

The output of an oscillograph.

What is a Toad?

*Y*ou may think that toads differ from frogs by being slow, short-legged, warty amphibians while frogs are smooth-skinned, long-legged, fast-moving, tailless amphibians. However, this is far from the truth. In fact, many frogs are slow-moving, short-legged and warty but they aren't all toads. In scientific terms, the only frogs that can be called 'true toads' are those that belong to Bufonidae family.

This family is distinguished from all other frog families by fine differences

The Cane Toad is the only true toad in Australia.

in bone structure. The only true toad in Australia is the Cane Toad, which was introduced into Queensland in 1935 with disastrous results.

Why are Some Australian Frogs Called Toads?

*A*lthough no true toads occur naturally in Australia, there are several species popularly called 'toads' or 'toadlets'. This kind of 'mislabelling' happens to all sorts of animals when people first arrive in a new country. When settlers from Europe arrived in Australia they saw certain species of frogs that looked rather like the toads they knew in their home country. So, not unnaturally, they assumed the Australian ones were toads too. Even though it turned out later that they were not true toads, the names were already in common use and so they remain today. These slow-moving, short-legged, warty Australian frogs are, however, all members of the ground frog family.

This might look like a toad — but it's a frog.

Sandhill Frog

2–4 cm

The Sandhill Frog doesn't just hibernate underground, it actually lives there.

Unlikely as it may seem, the coastal sandhills and dunes of Western Australia are where the appropriately named Sandhill Frog has made its home. These little frogs are cream to light brown with a patterning of darker brown and occasional red flecks. They spend their days buried beneath the sand, emerging at night to feed on ants. When they walk about the sand they leave distinctive marks that can be tracked back to where they burrowed into the sand.

Lifestyle

To cope with their unfriendly environment, these frogs form pairs in late winter or early spring and then go underground to mate. They remain beneath the sand for 5 or more months, moving deeper into the sand as the surface layers dry out. They may finish up more than half a metre below the surface — not bad going for such a small frog. Females lay 6–12 large eggs and the tadpoles develop underground within the egg. There is no free swimming stage at all. After several months the young emerge from their underground nursery to continue their parents' strange lifestyle.

> **DIVING IN HEAD FIRST**
>
> Sandhill and Turtle Frogs have an unusual method of burrowing into the coarse sand where they live. Unlike most burrowing frogs, they literally dive in head first. Burrowing into the sand at 45 degrees with their backsides in the air, they drag themselves underground with their arms, which are specially developed for digging.

Turtle Frog   5 cm

Looking like something from outer space (above) this frog burrows head first (right), unlike most burrowing frogs.

Another bizarre frog from Western Australia, the Turtle Frog looks more like a newborn turtle without a shell than a frog. It has a small head, stubby little arms, short legs and small eyes. Its colour ranges from grey, through yellow–brown to dark brown, and it has no distinct markings or patterns.

Diet and Reproduction

This frog feeds entirely on termites and is often found in termite mounds or in wood infested with termites. If it gets the chance, a Turtle Frog will overindulge on termites, eating more than 400 in one meal. Like Sandhill Frogs, male and female pairs burrow into the sand, sometimes more than a metre down. Here they mate and lay their eggs. Tadpoles develop entirely within their egg and emerge as fully formed frogs. The eggs, which can number between 30 and 40 in a clump, are the largest of all the Australian frogs' eggs and measure 5–7 mm in diameter.

Burrowing and Distribution

Like the Sandhill Frog, the Turtle Frog burrows head first, digging its way in with strong little arms. The Turtle Frog is fortunately quite abundant in parts of Western Australia and has not suffered the decline of so many other frog species.

What's so Bad About the Cane Toad Anyway?

Cane Toads will eat almost anything.

Cane Toads grow up to about 15 cm — much larger than most native frogs. They can live in almost any habitat and have no natural predators in Australia. They can also breed at most times of the year. As a result, they have spread alarmingly since their introduction in 1935 and have reached plague proportions in some areas. Their skin is highly poisonous and it is not known how many native animals may have died from attempting to eat them. They prey on native wildlife, such as lizards and mice and even other frogs, and may compete with native frog species for habitats and food. They live just as happily in suburban back gardens as they do anywhere else and have become a nuisance to residents as well as a hazard to pets.

What are Cane Toads Doing in Australia?

Cane Toads, which are a native of Central and South America, were brought to Australia in 1935 to control two native species of beetle that were a major pest to the sugar cane industry. In those days insecticide sprays were not available and the beetles had to be picked off by hand. The Cane Toad, which had been successfully used in Puerto Rico to control other species of beetle, seemed to be the answer to the cane-grower's problem.

The toads were first released at Cairns and Innisfail, and later at Mackay and Bundaberg. They were not successful in controlling the beetles, however. Firstly, the beetles were active when there was little growth in the fields and so the toads had nowhere to shelter while they supposedly went about their work of eradicating beetles. Worse still, these were flying beetles that seldom landed on the ground, whereas Cane Toads live on the ground and no one has ever per-suaded them to fly. So, the Cane Toads ate everything except the cane beetles.

The Cane Toad failed to control cane beetles.

BEETLE DRIVE
In the 1930s when cane beetles were such a major problem in Queensland, school children got up early in the morning to collect the beetles before school. They were paid sixpence (5 cents) a bucket for their work.

How Far has the Cane Toad Spread?

*T*he Cane Toad was brought to northern Queensland between 1935 and 1937. From its first release sites, it has spread, both north and south, all along eastern Queensland including most of Cape York. It has probably spread west as far as it can into the more arid lands of central and southern Queensland, but in northern Queensland it is still spreading west into the Northern Territory. It is estimated that the Cane Toad will have colonised the whole of the Top End of the Northern Territory by about 2020.

Cane Toads…on the march.

To the south the Cane Toad has moved down the New South Wales coast to somewhere near Coffs Harbour and is expected to continue to slowly spread further south. Cane Toads have been found breeding at sites near Newcastle and regularly turn up in Sydney suburbs.

How Poisonous is the Cane Toad ?

*T*he two large glands on either side of the Cane Toad's neck contain a toxic sticky fluid that may ooze or squirt out if the toad is alarmed or attacked.

A person is unlikely to die from merely touching a Cane Toad; however, if a Cane Toad's poison gets into the eyes it is extremely painful — although relief can be gained by bathing the eyes with plenty of water. Some people also react to the toxin if it gets onto cuts or abrasions, and some people have reported experiencing headaches after handling toads.

The effect on domestic animals, such as cats and dogs, who mouth or eat the toads depends on whether or not the toad releases venom. If it does, they have little chance of survival. This is also true for various native animals and birds that eat frogs, and there are many reports of deaths caused by eating or attempting to eat a toad.

Beware! A Cane Toad's neck glands hold poison.

Some animals can eat these creatures without being affected by the venom. These include several birds, rats, a freshwater turtle, crayfish, the Estuarine Crocodile and the Freshwater Snake. But obviously they take the toads only occasionally and cannot be regarded as major predators.

OTHER GROUND FROGS

Which is the Biggest Frog?

Australia's largest frog, the White-lipped Tree Frog.

*T*he largest known frog in the world is the massive Goliath Frog from west Africa. It can grow up to as much as 30 cm in length and weigh more than 3 kg. In comparison, Australia's largest frog — the White-lipped Tree Frog from the coastal regions of north-eastern Queensland — is a midget, growing only to about 14 cm.

The White-lipped Tree Frog is closely related to both the Green Tree Frog and the Magnificent Tree Frog, both of which grow to more than 10 cm. Almost as big as the Goliath Frog, the largest amphibian in Australia is the introduced Cane Toad which has been reported to reach 25 cm in length.

Which is the Smallest Frog?

*T*he smallest known frog in the world is a tiny 10 mm frog found in the Antarctic forests of Brazil. It is the smallest four-legged animal in the world. Australia's smallest frog is probably the Javelin Frog or the Rattling Frog, both of which measure only 14 mm in length and would fit comfortably on a five cent coin. The Javelin Frog belongs to the tree frog family and is found across the northern areas of Australia. The Rattling Frog is one of the narrow-mouthed frogs and occurs only in a small area of Cape York Peninsula.

FOREVER GROWING
The largest frog of any species is also the oldest, because frogs never stop growing — though their growth rate slows down after they reach sexual maturity. This is why scientists give only an approximate adult size for different frog species. So, when you see a very big Cane Toad or Green Tree Frog you know it has been around for a long time — at least by amphibian standards.

One of Australia's tiny frogs, the diminutive Javelin Frog.

Who's Bigger — Male or Female?

*F*or all but one species of Australian frog, the males are smaller than the females. There are certain advantages in this size difference. In the frog world bigger females produce more eggs, which boosts the species' chances of survival. When mating, the male grasps the female from behind and climbs either completely or partially onto her back. As the female often travels quite some distance while laying her eggs, the male may have to stay put for some time in order to fertilise the eggs.

The one species where the male is larger than the female is the Tusked Frog. This is possibly the case because males of this species fight each other, and size would be an advantage.

Most females are larger than males.

How Long do Frogs Live?

*A*lthough there are very few records around to tell us how long frogs live in the wild, we know something about how long they live in captivity.

With regular feeding and no predators, animals in captivity are likely to live a lot longer than those in the wild, but at least it gives us some idea of their life span.

The Green Tree Frog can live for more than 20 years given the right conditions, and many other species survive for 10–15 years. Even some of the smaller species have lived for more than 5 years in captivity. Records from overseas show that toads have lived for more than 30 years.

This big, old Green Tree Frog has been around a long time.

Red-crowned Toadlet

3 cm

This little frog displays a distinctive red 'crown' (above) and white leg- and armbands (right).

The Red-crowned Toadlet is found only in sandstone areas around Sydney, where it lives along small drainage lines (seepages) in the sandstone escarpment below a ridge. Unfortunately, the frogs' preferred habitat is also a popular site for humans and more and more houses are being built on top of cliffs or ridges to take advantage of the views. Sandstone areas not built upon are often degraded by stormwater pollution and the silting up of seepages. As a result, the Red-crowned Toadlet is now considered an endangered species.

Behaviour and Reproduction

This pretty frog gets its name from the bright red or orange patch on the top of its head. Males call all year round from burrows or concealed sites in litter. They often live in colonies and sometimes 10–12 individuals can be heard calling together. Females lay up to 20 large eggs in a depression or burrow and tadpoles develop within the egg until they are washed into pools during heavy rain. Tadpoles then have to complete their development quickly as these pools soon dry up. Male frogs usually stay with the eggs until they are flooded, but it is not known whether this increases the eggs' chances of survival.

> ### LET'S ALL CLAP HANDS
> Frog researchers and fauna surveyors, faced with the task of counting the number of Red-crowned Toadlets in an area get help from the frogs themselves. A few sharp claps or loud shouts usually prompt the males to begin calling, giving the researcher some idea of their numbers.

Corroboree Frog

3 cm

The Corroboree Frog calls the sphagnum bogs of the Snowies home.

There are no prizes for guessing how this easily identifiable frog got its name. Its bright yellow and black striped markings resemble the body painting used in some Aboriginal ceremonial dances, or corroborees. The Corroboree Frog lives in the Snowy Mountains and adjacent ranges at altitudes above 1000 metres. Sadly, this unique frog has recently suffered a serious decline in numbers in many areas of its range. No one is quite sure why.

Protected Life Cycle

Living at such altitudes, this frog is subject to extreme conditions of cold and snow for part of the year, but it has evolved a breeding strategy to protect its young. Males begin calling in mid-summer from concealed sites at the edges of seepages and sphagnum bogs. Mating occurs in spaghnum depressions and the female deposits 10–30 eggs inside the nest. The embryos develop within the egg and tadpoles emerge only when winter rains flood the nest or when melting snow raises the water level in spring. Tadpoles hatch and develop in pools. It is not until December or January that they metamorphose and leave the pools as frogs.

GLOBAL WARMING?
Corroboree Frog populations may have declined due to the recent increase in summer droughts in the Snowy Mountains area. Because of this frog's short life span and the low number of eggs it lays, several years of continuous bad weather could easily have reduced its numbers to a critical level.

Brown Toadlet
3 cm

Male Brown Toadlets call even during winter.

In the past this little frog was widespread and common in eastern and south-eastern Australia although, being hard to spot, it was not often seen. More recently its population has dramatically declined and the Brown Toadlet has disappeared from many previously known sites. No one knows the reason for its disappearance.

Also known as Bibron's Toadlet, this frog is distinguished by its black and white stomach and yellow patches at the base of each arm. Some also have a light stripe on their lower back and a bright yellow spot around their rear end.

The Brown Toadlet is usually found in grassland or forested areas in places where the soil is moist and there are ample hiding places, such as leaf litter and grass clumps. The male calls throughout the year and the female lays eggs under leaf litter or stones where they hatch when the site is flooded.

Sunset Frog
3.5 cm

The underside of the Sunset Frog is brightly coloured.

It is difficult to imagine how this brightly coloured frog could have escaped detection for so long, yet the Sunset Frog came to scientists' attention only in 1994 when it was discovered in the wetlands of southwest Western Australia.

The Sunset Frog is dark grey to black in colour above and has bright orange hands, feet, throat and chest, and a blue stomach mottled with white. It also has large parotoid glands behind the eyes.

The Sunset Frog has been found in only three locations, all of which are peat swamps. As these swamps are subject to fire, the frogs are at considerable risk, particularly in the drier seasons. Male Sunset Frogs call during spring and early summer and the females lay their eggs, one by one, into the pools or seepages in the peat swamp.

Common Eastern Froglet

3 cm

This little frog is common in the south-eastern part of Australia, including Tasmania. It has been found in high-altitude snow country on the mainland, but in Tasmania it seems to be confined to lower altitudes.

Common Eastern Froglets occur in all sorts of habitats including streams, ponds, seepages and roadside drains. They come in colours ranging from black through different shades of brown and grey. Even frogs living in the same area can

Common Eastern Froglets call by day as well as night.

vary in colour, pattern and skin texture. Some individuals are smooth-skinned, some are covered in small lumps, or tubercles, and others have raised folds of skin on their backs.

They call and breed throughout the year. Females lay up to 150 eggs in clumps in shallow water and tadpoles take about 8 weeks to complete their development.

Red-thighed Froglet

3.5 cm

As its common name suggests, this froglet has bright red patches on its thighs and groin. It has a large head with distinctive yellowish, or sometimes red, upper eyelids and it looks rather flattened. The red thigh and groin patches are the easiest way to identify this froglet as, like the Common Eastern Froglet, it comes in a variety of colours, patterns and skin textures.

The Red-thighed Froglet lays its eggs singly.

Red-thighed Froglets live in swampy areas, or in streams and gullies where winter rains provide ample water for breeding. Males call during winter and spring and females lay their large eggs separately in pools of shallow water. Despite the low temperatures, tadpoles metamorphose quickly, within one or two months.

The Red-thighed Froglet is also known as the Quacking Frog because of the sound the male makes when calling.

How Fast are Frogs Disappearing?

*F*rogs have been on Earth a great deal longer than humans and have survived all sorts of major changes to the Earth's surface and climate over millions of years. Unfortunately, their future survival does not look as good.

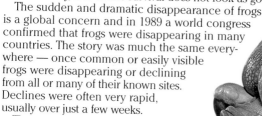

Clinging on for dear life.

The sudden and dramatic disappearance of frogs is a global concern and in 1989 a world congress confirmed that frogs were disappearing in many countries. The story was much the same everywhere — once common or easily visible frogs were disappearing or declining from all or many of their known sites. Declines were often very rapid, usually over just a few weeks.

The congress also noted that this significant decline started between 1978 and 1983 and the trend continues unabated.

Why are Frogs Disappearing?

*N*o one has yet been able to pinpoint exactly why frogs are disappearing so rapidly all over the world. In industrial Europe, many people blame acid rain, increased ultraviolet (UV) radiation caused by the destruction of the ozone layer and global warming. Killer fogs, unusual weather, and chemical pollutants have all also been blamed. Although some of these factors have probably played a part in the decline of some frogs at some localities, none can be seen as a common cause worldwide.

An exotic virus has been suggested as the more likely cause of the worldwide decline. Certainly

> **STRESS AND DISEASE**
> The fungus affecting frogs in some eastern States has spread to Perth, where five species are now infected. Frogs may be vulnerable to infection due to the stresses of environmental change.

some of the Australian species that have disappeared from Queensland's mountain rainforests appear to be affected by a virus. Several of these species are now regarded as extinct and many more from this area have suffered population declines.

Research done on these frogs and on frogs from similar habitats in central America has also identified a fungal infection on the skin of the frogs. Whether it is a fungus or a virus that is causing these declines in frog populations remains, at this stage, unknown.

A Great Barred Frog afflicted with a skin infection.

What's Going on in Queensland's Rainforests?

*O*ne of the most puzzling aspects of the decline of frogs around the world has been the sudden disappearance of frogs from isolated mountain areas, usually far removed from any pollution. The mountain regions of Queensland is just one of the regions that has suffered in this way. On the Conondale Ranges, for instance, the Gastric Brooding Frog and four other species either disappeared totally or their numbers declined sharply in 1979.

A Southern Day Frog — probably now extinct.

Since then, at least 14 Australian species of mountain rainforest frogs have disappeared or become rare. Some of these species also live in rainforest on the lowlands, and these populations have not been affected. Further, not all mountain rainforest species have experienced population decline. Most of those affected tend to be ones that live in a restricted range and lay small clutches of eggs. Those that appear unaffected have large clutches of eggs and are much more widespread.

How Does Pollution Affect Frogs?

*B*ecause frogs are continually moist, and absorb water through the skin, and because most frog eggs and tadpoles spend their time in water, the quality of the water in their habitats is extremely important.

Frogs and tadpoles are sensitive to a large number of waterborne pollutants. Apart from killing the frogs outright, some chemical pollutants affect the tadpoles' development and can cause deformities in bones and tissue.

These abnormalities become evident when the tadpole metamorphoses with missing or extra digits or limbs. There can also be less obvious deformities to the skull or the backbone . However, it is important to remember that abnormalities occur naturally in all species, including humans, so finding one deformed frog does not indicate pollution. An unnatural cause is suspected only when levels rise above the known normal level of abnormality.

A frog with facial abnormalities.

Hip-pocket Frog

2–3 cm

Males and females are often indistinguishable, but individuals can look different (above and right).

The most interesting thing about this little frog is the way it takes care of its tadpoles. Hip-pocket Frogs, also called Marsupial Frogs, live in Antarctic Beech forests and rainforest in the coastal ranges near the New South Wales–Queensland border where they hide beneath leaf litter, logs or other ground debris. Their grey to reddish brown back is usually marked with darker, often V-shaped, patches. A dark, broken line marks the change from the lighter coloured back to the dark grey or black of their sides. Most distinctive, however, are the male's two side pouches which he uses to hold developing tadpoles.

Pouch Development

These frogs do not need free water in order to breed. Instead, the female lays her eggs on the ground and during the 10–14 days they take to hatch, both the male and female remain guarding the eggs, possibly to prevent beetles attacking them. As soon as the tadpoles begin hatching, the male takes over their care. He climbs amongst the eggs, getting covered in the jelly so the tadpoles can swim and slide up his body and wriggle into the pockets of loose skin in his sides. Although between 8 and 18 eggs are laid, only about half make it to the pouches. The others die within a few days. Those that do make it, stay there, sustained by the egg yolk for about 7–10 weeks. They leave the pouch as fully developed frogs at about 4 mm in length.

While the tadpoles are in his pouches, the male goes about his normal daily activities with the added burden of a squirming bunch of babies in his pockets.

Tusked Frog

4.5 cm

The Tusked Frog is easily identified by its bright red markings in the groin area.

The Tusked Frog is one of only a few Australian frog species where the male is larger than the female. In fact, the male's body is only slightly larger than his head. Both the male and female Tusked Frog have a pair of tusks at the front of their bottom jaw, but the male's tusks are larger. As these tusks fit into special pits in the roof of the mouth, they do not protrude outside the mouth and so are seen only when the frog opens its mouth.

Description

With its mottled, dark brown back, the Tusked Frog appears to be nothing special when viewed from above. Its underparts, however, are stark black and white with bright red patches in the groin area and on the legs. Individuals of this species have their own distinct black-and-white markings and no two frogs exhibit the same patterning, so individuals are easy to tell apart.

Reproduction

The male Tusked Frog is territorial and in combat uses its tusks to bite other males around the throat. The male with the biggest head and longest tusks probably has quite an advantage.

The female of this species lays several hundred eggs in a floating foam nest and the male remains close to the nest for a few days until the tadpoles have hatched.

DINING OUT ON LOCAL FOOD

Tusked Frog females favour grasshoppers, crickets and similar insects, while males prefer pond snails. This difference in food preference is probably because males spend a lot of time around ponds while females are more land-based.

Who Preys on Frogs?

*F*rogs are a favourite food for a wide range of animals. Even some people are partial to eating frogs, or at least parts of them. Many aquatic animals, such as crocodiles, fresh-water turtles, water rats, water beetles, other frogs and even certain spiders, include frogs or tadpoles in their diet. Some water birds are also major predators and so are several species of snakes and lizards and, of course, fish. In Australia the Ghost Bat also catches and eats frogs. Newly metamorphosed frogs are susceptible to attacks by insects and ants, which makes their transition from water to land hazardous. The so-called Mosquito Fish (*Gambusia holbrooki*), introduced into Australia to eat mosquito larvae, possibly does eat some mosquito larvae but sadly it also eats frogs' eggs and tadpoles, too.

A Carpet Python putting the squeeze on.

How do Frogs Protect Themselves?

A Giant Banjo Frog blends into the leaf litter.

MONSTROUS DISGUISE

Before the Cane Toad uses its last line of defence — the toxic substance it expels from glands in its head — it attempts to frighten off its enemy. It does this by inflating its body, raising itself clear off the ground and tilting its body towards the threat. With this bluff, the toad no doubt hopes to convince the predator that it is far too big and dangerous to be eaten.

---

*B*eing such a desirable food item for so many other animals means that frogs need exceptionally good methods of avoiding their enemies.

A leap has taken many a frog well away from the jaws of death. Leaping into water is a particularly successful survival strategy as the frog can then swim away underwater.

Camouflage is another method used by many frogs and among the best examples of this are the Green Tree Frogs. When they flatten themselves against the surface, tuck in their limbs and stay perfectly still, they are almost invisible in bushes, trees or reeds. For other frogs, simply remaining hidden during the day keeps them out of sight of enemies. Quite a number of frogs protect themselves by releasing toxic, distasteful or irritating substances from the glands in their skin when alarmed.

Why are Some Frogs so Bright?

*N*ot all frogs take on camouflage colours to blend in with their background and so remain hidden. Some frogs are brightly coloured and can be picked out very easily from their surroundings. Some of these colourful frogs secrete toxic substances when threatened or attacked and their colouration probably serves as a warning to potential predators to keep well away or risk being hurt.

Bright flash colours such as those on the legs of a Peron's Tree Frog may startle a predator and discourage pursuit.

Even among well-camouflaged frogs, there are often hidden spots on the groin area, stomach or rear end that are brightly coloured and as the frog jumps away the bright colours flash out, startling the predator and discouraging it from further pursuit.

Do Frogs Shed Their Skin?

*A*ll frogs shed their skin at intervals which range from a few days to several weeks. Skin shedding is not the protracted affair it is with snakes. Nor is it a time when frogs are particularly vulnerable, as are snakes.

Frogs may replace their skin hundreds of times in their lifetime. In this process, called moulting, the outer layer of skin separates from the layers beneath to expose the new skin. The frog works at removing the old skin by flexing various muscles, arching its back and scraping at its body and head with its limbs to separate the outer skin from the new one underneath. The old skin normally comes away in one piece and the frog then stuffs the torn edges into its mouth and proceeds to eat it.

When burrowing frogs emerge above ground after a long spell below they receive immediate nourishment by eating the cocoon that was formed by the succession of shed skins while they were underground.

A moulting frog pulls its old skin off.

Smooth Toadlet 3 cm

Although the common name for this frog is the Smooth Toadlet, its skin has, in reality, a rough, warty appearance due to the many raised tubercles on its back. This frog has a pale triangular mark on the top of its head and red to orange patches in its groin and at the back of its thighs.

Its camouflage-colouring makes this frog hard to find.

The Smooth Toadlet is widespread throughout south-eastern Queensland and eastern New South Wales, and its distribution extends into Victoria. It can be found in forests and woodlands, often within grassy areas that have become flooded after rain.

The male Smooth Toadlet calls through spring and early summer from the edges of these temporarily flooded grasslands. Because of its colouring matching the background and its habit of hiding beneath leaf litter or in burrows, this frog can be difficult to find, even when it is calling. Despite its broad distribution, no scientific study has ever been done on this frog. As little is known about its behaviour or life history, it remains a fine candidate for further investigation.

Fletcher's Frog 5.5 cm

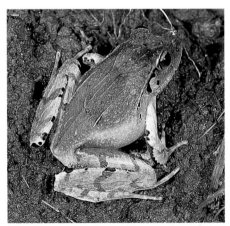

Fletcher's Frog is also known as the Sandpaper Frog because of the rough skin it develops during the breeding season. This frog is a forest dweller and is found in rainforests and in other wet forest types. It lives on the ground in leaf litter or in tree hollows, from where it emerges during rain in the summer months. Despite its moderate size, its colour and patterning make it very hard to find as it blends in so well with the leaf litter.

Fletcher's Frog is well camouflaged among leaf litter.

Breeding occurs during summer. The female lays up to 300 eggs in a foam nest. The tadpoles have a reputation for eating each other and only the fittest survive. They also eat the eggs and tadpoles of other frog species.

Nicholls' Toadlets sometimes live in bull ant nests.

Nicholls' Toadlet   2 cm

Although a close relative of the burrowing Turtle Frog and Sandhill Frog which inhabit arid- to semi-arid areas, the Nicholls' Toadlet lives in the high-rainfall area of south-western Western Australia.

The upper parts of this frog are a drab, dark brown, the underneath is mottled blue and white and there are bright orange spots at the base of its arms and legs. Like most of the toadlets, this frog doesn't look as if it would be capable of climbing, but males have been observed on rainy nights perched up to 50 cm above the ground in bushes, calling vigorously. Why they sometimes climb to call is a mystery, as they normally call from the ground. The eggs of this species are large and the tadpoles develop entirely within the egg.

Several of these frogs have been found inside the nests of bull ants without any apparent hostile response from the resident ants. This is probably not normal behaviour for these frogs as they are usually found in and beneath rotting logs.

The Karri Frog is confined to the wet Karri forests.

Karri Frog   2.5 cm

The Karri Frog also goes by the name of the Roseate Frog. It is a small frog with a brown back and pink stomach. The male's throat is black, while the female's is pink — the same colour as her stomach.

This frog is only known to inhabit a small area that extends no further than 70 km between Margaret River and Walpole in Western Australia. The Karri Frog inhabits drainage lines in swamps and the female makes use of rotting logs to make her burrow in which she lays her eggs. A clutch contains an average of 24 eggs, and tadpoles remain in the burrow, relying on the egg yolk to sustain them until they metamorphose.

The Karri Frog belongs to a group of four frog species that share the common name of Ticking Frog. This name describes the male's call, which sounds like a series of rapid clicks or ticks. All four species are restricted to living in small areas in the extreme south-west of Western Australia.

Will Frogs Like my Garden?

Four species of frog call from this small pond in Sydney.

*I*t is not difficult to attract frogs to your garden as long as you can provide the right kind of water, shelter and food.

First of all, you need a permanent source of water. It doesn't have to be a large area, nor does it have to be very deep. Enthusiasts have created very successful frog habitats out of nothing more than a child's wading pool.

Next, the frogs must have somewhere to shelter — this means lots of plants and ground cover around the water. A boggy zone about a metre wide at the edge of the water is also a good idea. Last, but not least, frogs need food. A garden with plenty of plants, leaf mulch and ground cover will normally have plenty of insect life, which is exactly what the frogs need.

Place your frog zone where it will be sheltered from the wind, and where it will be partially shaded — about 50 per cent sun/shade is ideal.

Will Frogs Stay in my Garden?

A frog-friendly habitat.

*H*aving attracted some frogs to your garden, or perhaps having introduced them from somewhere else, you need to maintain a frog-friendly environment so they want to stay. This comes back to the three basic factors — water, shelter and food. Frogs may come to your garden because you have a pond, but they won't stay if there is not enough shelter. Frog heaven is not manicured lawns, trimmed shrubs and everything neat and tidy. Frogs need places to hide, particularly during the day. You should keep pesticides and fertilisers well away from your frog zone so they won't pollute the pond water. Dogs, and particularly cats, are not friendly to frogs. Keep cats inside at night when the frogs are active and fence the area if you have dogs.

Will Frogs Breed in my Garden?

*I*f you have created the right kind of environment for frogs, then species that breed in still or slowly flowing water should be happy to breed in your pond. That is, as long as the water is unpolluted.

Tap water contains chemicals, such as chlorine and fluoride, so you should let it stand in the pond for at least a week before introducing the animals. Alternatively, you can buy a chemical from the pet shop which will take the harmful additives out of the water more quickly.

If you want your frogs to breed, don't expose your pond to full sun. The water will become warm, thereby reducing the oxygen available to growing tadpoles and certainly encouraging algae growth. Although tadpoles eat algae, too much algae will affect the water quality. Too much shade, on the other hand, inhibits the growth of algae and other vegetation.

> **BEWARE!**
> **FROGS CALLING!**
> Don't put your pond too close to your bedrooms unless you enjoy the sound of many frogs calling loudly all through the night. Also, be kind to your neighbours and keep your pond away from their bedrooms, too.

Frogs love wet places.

What About Mosquitoes?

*M*osquitoes are attracted to bodies of still water and so they will be attracted to your frog pond. There are several ways to control them, however.

Firstly, small frogs will eat some of the mosquitoes. A healthy pond, apart from encouraging frogs, will support a variety of insects, some of which will eat mosquitoes. Dragonflies, for example, eat both mosquitoes and their larvae.

As mosquitoes lay their eggs in still water, a circulating pump will cause enough movement to discourage many of them. A pump will also help oxygenate and maintain the quality of the water.

The best mosquito controllers of all are probably small fish, but be very selective about these. Under no circumstances put in the misnamed Mosquito Fish because they eat frogs' eggs and tadpoles. There are several species of native fish that are quite suitable and you need only three or four in a small pond.

*Mosquito Fish (*Gambusia holbrooki*) eat frogs' eggs and tadpoles, so don't put them in your frog pond.*

Baw Baw Frog

5 cm

The Baw Baw Frog is threatened by skiers.

This frog lives only above 1100 m on Mount Baw Baw, in Victoria. It is considered endangered due to its restricted distribution and the threat to its habitat posed by recreational activities.

The Baw Baw Frog is dark brown with a speckling of lighter and darker flecks. The glands at the back of its head are darker and prominent. It has yellow markings between the eyes and these may extend down the back.

This frog colonises areas near seepage pools in wet heath or bogs and breeds in midsummer. The female lays 50–150 large eggs in foam in a depression beneath ground cover where the soil is damp. Unlike the mountain-dwelling Corroboree Frog, whose tadpoles take almost 12 months to develop, the Baw Baw's tadpoles metamorphose within 5–10 weeks, relying on their large yolk reserves as they develop into frogs.

Sphagnum Frog

3.5 cm

Sphagnum Frogs prefer really soggy areas.

The Sphagnum Frog is so-named because the first specimens were found living in sphagnum moss. Its distribution is restricted to areas of high rainfall on the eastern side of the Great Dividing Range in northern New South Wales. The Sphagnum Frog is normally found in high-altitude rainforest close to the headwaters of creeks or seepages where the ground is well saturated. Males are greyish green on the back and females are reddish brown.

This frog breeds during the summer months. The male frog calls for a mate from his water-filled burrow or from some other concealed position. The female lays a clutch of 40–60 eggs in a frothy mass at these calling sites and the tadpoles stay at the site after hatching. The tadpoles take approximately one month to complete their metamorphosis into frogs. During this time they survive solely on the yolk of their egg.

Yellow-bellied Mountain Frog

3 cm

This spectacular-looking frog has a brilliant yellow-coloured stomach, while its back is mottled and features colours ranging from bright red through to almost black. It was first described as recently as 1975, and was assigned the scientific name *Philoria kundagungan*; 'kundagungan' is an Aboriginal word meaning 'mountain frog'.

This species lives only in mountainous rainforests.

The Yellow-bellied Mountain Frog appears to have a restricted distribution. It has been found in only eight locations on the ranges of the Queensland–New South Wales border. It lives in very soggy leaf litter or in mud along small creeks in mountainous rainforests.

Like the moisture-loving Sphagnum Frog, it lays large eggs in water-filled depressions or burrows and its tadpoles survive their period of metamorphosis sustained entirely on the reserves stored in their egg yolk.

Carpenter Frog

6 cm

Found only in the rocky escarpments of the Kimberleys and Arnhem Land, this frog has a large, conspicuous outer ear covering that is about the same size as its eye.

The Carpenter Frog is coloured grey–brown to dark brown, often with many lighter or darker blotches. It was given the name Carpenter Frog because the male's call is a single short tap, similar to the sound made by a hammer striking a piece of wood.

The large ear-covering is a distinctive feature.

This frog lives in caves during the dry season and leaves its retreat when the wet season arrives. It then shelters under flat rocks at the edge of creeks, from where the male begins to call during summer. The female deposits a few hundred eggs in a foamy nest into rock pools on the creek bed. The tadpoles are black and may reach 5.5 cm in length before metamorphosing when about 2–3 months old.

Frogs: Fact or Fiction?

Is there a prince hiding under that frog skin?

*F*rogs have attracted a lot of bad publicity and bizarre behaviour through the centuries, probably because they were considered ugly, slimy creatures that no one could possibly love. Perhaps people also saw something magical in the way they changed form from tadpole to frog. Metamorphosis may well have been the inspiration for the story of the princess who kissed the frog which then turned into a handsome prince.

Frogs, or parts of them, have always been an essential ingredient in any self-respecting witch's brew. They have also been used by non-witches in medical 'cures'. Whooping cough was said to be cured by placing a dead frog in a box and tying it around the patient's neck until it decayed. Another folk remedy recommended placing a dried frog in a silk bag and hanging it around the sufferer's neck to cure epilepsy and 'other fits'.

Are Frogs Useful to Humans?

*I*n some parts of the world frogs are eaten by people. This is the case in New Guinea, for example, where some tribes hunt frogs and have devised a range of cooking methods for them. In some parts of Australia, too, traditional Aborigines occasionally eat frogs.

In Europe, frogs' legs have been a delicacy for centuries. Even Mrs Beeton's *Book of Household Management*, the English cooking 'bible' for many years, supplied a recipe for stewed frogs.

Apart from their use as food, frogs have long been used for dissection purposes as a teaching aid for students of biology. They have also been used in research and in experimental studies. In recent years considerable attention has been given to the substances found in the skin glands of various frogs. Some of these substances have been found to have medicinal properties and further important research may well result in useful products in the future.

Until the introduction of modern pregnancy test kits, the Cane Toad was used in laboratories to confirm pregnancy in humans. Their skins have also been used in bookbinding, and in the manufacture of a range of bizarre tourist ornaments.

A purse made from a Cane Toad's skin.

What's so Special About Frog Skin?

*W*hat makes the frog's skin unique are the glands in it and the way liquids and gases can pass through it.

Frogs have several kinds of glands in their skin. One type of gland is responsible for secreting a fluid that keeps the frog's skin moist. Other glands secrete a fluid to deter potential predators by being either poisonous or very unpleasant to taste. Another group of glands secrete antibiotic substances that help prevent skin infections.

Frogs can reduce water loss by huddling together.

Instead of drinking water, as most animals do, a frog takes in water through its skin. Its skin can also absorb oxygen and expel carbon dioxide — this is a very useful process for a frog that is dormant during cold weather because breathing through the lungs at this time would required precious energy.

A frog's skin also contains pigments that are responsible for the frog's colour. Some frogs can change their skin colour by rearranging the pigment granules within the cells. This change may occur as a response to changes in temperature or light intensity, or because the animal is stressed.

What's the Best Way to Keep Tadpoles?

*F*irst you have to catch them. An aquarium net or a strainer works well to scoop them out of the water. Don't take too many tadpoles at a time as they won't develop if they are overcrowded. Plastic bags are ideal for transporting them but whatever you use, try to prevent the water and tadpoles from slopping around too much.

Tadpoles can be kept in plastic or glass containers, or in aquariums. If you are using tap water you must remove the harmful chemicals from it before you put in the tadpoles. You can buy a product from a pet shop to do this.

Feed young tadpoles with small pieces of lightly boiled lettuce and, as they grow, add some fish food. Only put in enough food for them to eat quickly, or the water will become foul and need changing. Provide a sloping piece of rock or other material so that the metamorphosing tadpoles can clamber up from the water's edge. Cover the top of the container with gauze to stop the frogs leaping out.

Tadpoles can be kept in a shallow dish, glass container or aquarium.

Spotted Grass Frog

3–5 cm

This frog is a successful coloniser in new areas where semi-permanent water bodies form.

The Spotted Grass Frog is widespread over south-eastern Australia and is one of the most common frogs in this area. Surprisingly, a colony of this species has become established at Kununurra in the Kimberley region of Western Australia, almost certainly by accident, possibly by being brought in with plants or transportable houses from South Australia.

This frog has a grey–green back splotched with brown or olive green and has bars of the same colour on its limbs. Some frogs of this species have a pale stripe running down the middle of their back. Breeding males have a yellow throat, making them easily visible during calling when inflating their vocal sac.

Habitat

Usually found in swampy areas and around ponds and dams, Spotted Grass Frogs will occasionally take up residence in garden ponds. Males often head for shallow water in flooded paddocks or in drains and ditches next to roads when calling. Breeding takes place from spring to autumn and females lay eggs in a nest of foam on the water.

Calling

This species has three distinct 'call races'. Males living in the northern part of the range have a rapid clicking call. Those to the south make a single sharp click, and those in western parts fall somewhere in between, making two or three rapid clicks in succession.

> **TRUCKIN' FROGS**
> Some frogs have colonised areas a long way from their usual homes by being transported along with plant or building material to faraway places.

Great Barred Frog

10 cm

The handsomely marked Great Barred Frog is confined to the edges of permanent streams in rainforests.

This frog is one of five species of barred frogs. It is the largest of the Australian ground frogs and has distinctive bars on its powerful back legs. The Great Barred Frog lives in the rainforests and wet sclerophyll forests of the east coast of New South Wales and southern Queensland. It hides during the day in leaf litter near permanent fast-flowing streams and emerges at night to forage in the litter or along the banks of a stream.

Usually tan to copper-coloured on its back, this large frog has a dark stripe running from its snout, through each eye and over its ears. It also has a conspicuous row of black spots on its sides.

Reproduction

Mating and egg laying occur in water from late spring to early summer. The female lays just a few eggs at a time and, after they've been fertilised, she catches them on her foot and tosses them out of the water onto the creek bank where they land on grass and other vegetation. Because the eggs have a sticky coating they stick easily to leaves and twigs. After hatching the tadpoles either fall into the water or are washed in during rain. Tadpoles can grow to more than 7 cm long.

73

The call of the Northern Bullfrog is a short 'bonk'.

Northern Bullfrog 7 cm

The Northern Bullfrog is also known as the Northern Pobblebonk or Northern Banjo Frog. This frog has an impressive build and is strikingly coloured. It has a stocky body and its back is dark brown or occasionally black. The gland below its eyes is a prominent vivid yellow and the stripe along its sides is orange. Its throat and stomach are also yellow while the groin and back of the thigh are red and yellow.

The Northern Bullfrog occurs right along the east coast of Queensland and its distribution extends inland through the western slopes of New South Wales. This frog inhabits areas of woodland where it congregates near bodies of permanent water, such as dams and large ponds where it hides in vegetation.

Males call throughout summer and autumn, usually from holes at water level. Females lay their eggs in a frothy foam nest attached to vegetation.

The Marbled Frog is a northern Australian species.

Marbled Frog 6 cm

The Marbled Frog can be found throughout northern Australia and also occurs in New Guinea. This frog has a grey to olive back marbled with dark blotches. These blotches are raised and wart-like in form and a male may also have some sharp, black spines on its back. It has a prominent white or pale yellow fold of skin running from below each eye to each arm.

This frog normally lives in swampy areas where it conceals itself in long grass or other dense vegetation. The male calls during spring and summer while remaining hidden in the vegetation. The Marbled Frog tends to be solitary and does not congregate in large numbers at suitable breeding spots as many other species do. Females lay eggs in a foam nest on the top of the water. Tadpoles are black and reach 7 cm before metamorphosing.

Striped Marsh Frog 7 cm

The Striped March Frog is commonly a casualty in suburban swimming pools.

A common inhabitant of suburban ornamental garden and fish ponds, this frog calls continuously after heavy rain throughout most of the year — often to the annoyance of its human neighbours. Its call resembles a single loud 'plunk' or 'whuck' sound.

The Striped Marsh Frog is distributed throughout eastern Australia, extending to Tasmania. It lives near bodies of permanent, still water as well as in suburban gardens and hides under debris during the day. It has a strikingly patterned back, striped in shades of light and dark brown with often a pale stripe running down the middle of its back. The male has a yellowish throat with a darker flecking.

Breeding occurs throughout the warmer months and females usually lay up to 1000 eggs. Except in South Australia and Victoria, eggs are laid in a frothy nest.

Salmon-striped Frog 7.5 cm

This frog is seen only in the breeding season.

The Salmon-striped Frog is easily identified by its distinctive colourful patterning. It is brownish–grey on its upper body with scattered dark brown blotches and three pink to orange–red stripes on its back. It has similar colouring on its arms and legs, and a further orange–red stripe runs from below each eye to the top of the arms. The frog's stomach is white but its groin and the thighs are a marbled black and white.

This frog can be found throughout south and eastern Queensland and its distribution extends down to north-eastern New South Wales where it occurs inland across to the central western areas of the State.

In the breeding season, from spring through to autumn, these frogs are often found during the day sheltering under rocks, wood or debris at the edge of swamps and ponds. At night males call from vegetation and debris at the water's edge. During the rest of the year they remain under the ground.

Can I Keep a Frog in Captivity?

Whether or not you will be able to keep a frog in captivity depends on the laws in your state. Frogs are totally protected in most states, so it is absolutely vital that you check first with your state fauna authority to find out exactly what you can or cannot do. Once you have established the legal position, you need to ask yourself if you are capable of keeping a frog. Frogs have particular requirements that you will

Froggie friends need clean water, shelter and food.

have to meet if you intend to keep them in captivity. One of their most important requirements is food. Frogs do not eat cornflakes or vegemite; they are carnivores and eat other animals, mainly live invertebrates. You need to make sure you know of a reliable supply of goodies such as crickets, moths, cockroaches, grasshoppers and beetles, for your frog.

How do I Get a Frog?

Again, this will depend on the laws in your state. If there are no restrictions, then getting a frog is easy — just go out and catch one. (More about where and how on pages 90 and 91.) If there are restrictions you will probably only be able to get a frog from someone who already keeps and breeds frogs. If you don't know anyone like this, then the best approach is to

Peron's Tree Frog, an accomplished escape artist.

contact a frog-interest group in your area. Your state conservation authority can probably give you information or a contact phone number. Some of these groups are listed at the back of this book.

RESPONSIBLE RETURNS
What do you do if you no longer want to keep your frog? You can release it, but only back into the area from which it came. Releasing organisms into new areas can cause all sorts of changes, many of which are detrimental to the lives of the original inhabitants. If you don't know where the frog came from, or cannot get there, do not release it. Give it to someone else who is interested in keeping a frog. If you don't know anyone, contact your nearest frog-interest group.

How do I Look After my Captive Frog?

*Y*ou'll get the best advice by joining a frog-interest group. Most members of these groups keep frogs themselves and they know exactly what they need. Warm-climate frogs won't tolerate cold conditions, so an aquarium inside is the first requirement. For tree frogs, the sides of the aquarium should be high so there is plenty of vertical space. Terrestrial frogs need more floor space. The aquarium should have a well-fitting lid with fly-screen ventilation. Keep the aquarium in a warm spot but

An ideal set up for Green Tree Frogs.

away from direct sunlight. Provide a bowl of clean water and hiding places. If you have a tree frog, provide branches for climbing. If it is a terrestrial frog, provide sand deep enough to burrow into. Most importantly, find out about your frog's requirements before you get it.

Can I Keep my Frog Indoors?

*F*rogs are ideal animals to keep indoors, although you will need some form of subdued lighting to maintain a warm and humid environment. Red or blue lighting is good for this purpose. If you intend to landscape the aquarium with plants, then you should also install a fluorescent plant light. Frogs are normally active at night so don't place the aquarium in bright light. If you intend to keep frogs indoors, think about how you will clean the aquarium and replace materials from time to time. Remember also that male frogs will call and they do this with greater intensity at night. Light sleepers in the family should be consulted.

A frog's aquarium is its castle.

Southern Gastric Brooding Frog   5 cm

The Southern Gastric Brooding Frog spends most of its time in the water and is very slimy to touch.

This frog is one of only two in the world that breeds its young in its stomach. Although no one has ever seen it happening, the female apparently lays large eggs and then, after fertilisation, swallows them. The tadpoles are nourished during their development by their large egg yolk. At about 6–8 weeks the mother comes to the surface of the water, opens her mouth and the young frogs are 'born', one or two at a time, over several days. The tiny frogs sit on their mother's tongue and then jump out into the water.

Description and Behaviour

The Southern Gastric Brooding Frog is dark grey to brown on its back and white to yellow on its stomach. It has many glands under its skin. This frog is a strong swimmer and lives entirely in the water, often hiding under stones in creek beds. Sometimes it just drifts about, occasionally even floating on its back.

Discovery and Decline

The Southern Gastric Brooding Frog first came to the attention of scientists in 1972 when it was discovered in rainforest creeks in the ranges of southern Queensland. In 1984 a second, similar species was found in central coastal Queensland. Sadly, both species have since disappeared and the Southern Gastric Brooding Frog is probably extinct.

LIFE ON THE INSIDE
While the young are maturing, they are protected from harmful gastric juices in their mother's stomach by a secretion that stops the production of stomach acids. While the young develop, the mother doesn't eat.

Sharp-nosed Torrent Frog

3 cm

The Sharp-nosed Torrent Frog may be teetering on the brink of extinction.

This small frog inhabits the cool, wet rainforests of Queensland above 300 m where it lives around fast-flowing streams. It is active during the day and will sometimes bask in sunlight. This frog has a very pointed snout that projects beyond its lower jaw. Its back is light brown to orange–brown and its sides are dark brown to black with a distinctive skin fold separating the colours. There is a conspicuous white patch at the base of each arm, and the groin and underside of the limbs are usually yellow.

Behaviour and Reproduction

The male always stays close to streams where it calls from among rocks or litter. Its call is a sharp-pitched tapping sound, which is why this frog is sometimes called the Tinker Frog. The female lays 25–40 large eggs in a clump in the water among rocks. Tadpoles have large, rounded mouths which they use to hang onto rocks to avoid being swept away by the fast-flowing streams.

Conservation Status

This is another species of frog whose numbers have declined severely; in fact, this frog was thought to be extinct until recently when two sightings were made. Whether it is making a comeback or these are the last of a doomed species, no one knows.

HERE I AM
The white marks on the throat and chin of the male frog are noticeable only when it is calling and its vocal sac is inflated. These marks might help to attract females, or perhaps they are there to warn off rival males.

NARROW-MOUTHED
AND TRUE FROGS

Who are the Narrow-mouthed Frogs?

Fry's Frog is one of 16 narrow-mouthed frogs in Australia.

*N*arrow-mouthed frogs form a family of frogs that is plentiful in Africa, Asia, South America and New Guinea. In New Guinea they are very widespread with about 100 known species. Australia is home to only 16 species. These small, secretive frogs are confined to the far north of Australia, and most of them inhabit only moist rainforests.

Although these frogs live mainly on the ground, some have discs on their fingers and toes and have been found in bushes or in plants growing on larger trees. They get their common name from the fact that some species have a narrow, pointed head. Narrow-mouthed frogs don't have very powerful legs and are walkers rather than leapers. As far as researchers know, they all lay a small number of large eggs on the ground and the tadpoles develop within the eggs. Whenever egg clumps have been found, there has always been a male frog close by.

Who are the True Frogs?

The Water Frog is the only true frog representative found in Australia.

*T*his family of frogs has a wider habitat range than any other family and is found in Europe, the Americas, Africa, Asia, some Pacific islands including Fiji, and in New Guinea. Only one species is found in Australia, however, and it is confined to northern Queensland and the Northern Territory. True frogs have long legs, can jump well, and live at the edge of water.

They have been used in research and teaching for many years and much of what we know about frogs comes from these studies. True frogs are also the ones that provide the main ingredient for the frogs' leg dishes featured on some restaurant menus. Because they are so well known, many people have come to think of their appearance and habits as the norm, and so they have acquired the name of 'true frog', although the group is no more truly frog-like than any other group.

WHAT A MOUTHFUL!
The largest frog in the world, the Goliath Frog from west Africa, belongs to the family of true frogs and can grow 30 cm long. Luckily for the Goliath Frog, no one seems to have thought of adding jumbo frogs' legs to the menu — yet.

How Well can Frogs Hear?

*M*ale frogs call to attract females, to mark out their territory and to warn off intruding males, so other frogs need to be able to hear these calls clearly and interpret them. Frogs can tell the difference between the calls of their own species and those of others, even when there are six or seven different frog species calling around the same pond. It appears that they can do this because they are only 'tuned in' to the call of their own kind, while the calls of other species are

The black hole to the left of this frog's eye is its external ear membrane.

blocked out. Small frogs may only be able to pick up sounds a few metres away but larger species can hear calls over a distance of 100 m or more.

How Well can Frogs See?

*M*ost frogs will only eat prey that is still alive and wriggling, so frogs are attracted to their food source by its movement. For this reason frogs need very good eyesight and they generally have large, sticking-out eyes. They do have one blind spot, however, which is straight in front of their snout. Because of this blind spot, frogs have to turn their head to one side when prey appears right in front of them. The majority of frogs are more active during the night than they are in the daytime so their night vision is generally very well developed. There is no evidence to suggest that frogs can see colours.

A frog cannot see directly in front of its snout.

Northern Territory Frog   2 cm

The Northern Territory Frog may climb to call.

This little frog lives in the Top End of the Northern Territory and adjacent islands. On the mainland it is found around undisturbed areas in permanent swamps and streams. On Melville Island, on the other hand, it is commonly seen on roads and on lawns. It usually calls while remaining hidden beneath leaf litter and other debris, but frogs have been seen calling from vegetation up to 50 cm off the ground.

This frog is grey or light brown with darker speckles which are more numerous on its head. There is usually a narrow stripe down the middle of its back. Although geographically separated, it is identical to the Slender Frog found on Cape York and in New Guinea, and can only be distinguished by its call. These two species are referred to as sibling species because they share a common ancestor.

Rain Frog   3 cm

The Rain Frog's call is a very high-pitched whistle.

The Rain Frog is also known as the Whistling Frog and the White-browed Whistling Frog because of the male frog's 'whistling' mating call. This species inhabits areas of rainforest in northern Queensland between Cooktown and Ingham, living at both low and high altitudes. It hides under rocks and logs, as well as in leaf litter. The male Rain Frog calls from its hiding place in the leaf litter in a series of rapid, high-pitched whistles.

This frog is pale to dark brown in colour and is patterned with darker mottled markings. The areas around the frog's chin and throat are usually coloured purple. It has a white stripe on its upper eyelids and its eyes are bright red.

Scientists know little about the Rain Frog except that the female probably lays its eggs on land and tadpoles probably develop inside the egg.

Robust Frog
3 cm

This species lives in the up-
land rainforests between
Townsville and Cairns and
has never been found in
the lowland rainforests of
that region, although it is
quite common in its area.
It is often found beneath
rocks and logs. It ranges
from reddish-brown to dark
brown in colour with darker markings, especially on
the side of its head.

The Robust Frog lives in high-altitude rainforest..

The male calls from
under the leaf litter next
to his shelter in short, high-
pitched chirps. Although both the Rain Frog and Fry's Frog live in the same
area and look similar to the Robust Frog, its call can be distinguished from
the calls of these other frogs by its softer, higher and more musical quality.

The female lays her eggs directly into the leaf litter on the forest floor, and
the tadpoles complete their development entirely within the egg.

Ornate Frog
3 cm

The Ornate Frog is wide-
spread in the rainforests of
northern Queensland and
occupies habitats ranging
in altitude from sea level to
the mountain forests.
Although it usually lives on
the ground, it has expanded
pads on its fingers and toes
and is a good climber. The
male sometimes climbs up
to 2 m above the ground to
call. Its call, which can be
heard on warm, wet nights,
is a short, bleating sound
and is surprisingly loud for
such a small frog.

Male Ornate Frogs stay with the eggs as they develop.

The female lays between 10 and 20 large eggs on the ground, but out
of sight. The male stays with the developing eggs for at least some of the time.
The tadpoles are sustained entirely by the egg yolk and emerge from the egg
as fully developed frogs without ever having been in the water.

Rock Frog

4.5 cm

The Rock Frog is found among granite boulders in only one locality.

The Rock Frog is the largest of the Australian narrow-mouthed frogs. It is the only species in the group that doesn't live in rainforests. Instead, it lives among the lichen-covered granite boulders of Black Mountain near Cooktown, in Queensland. Its habitat contains very little vegetation except for an occasional tree but the granite boulders provide many hiding places, not just for frogs but for other animals too. In some places crevices in the boulders extend into the mountain for 10 m or so and are wide enough for an adult human to climb into. These crevices are an ideal environment for frogs because of the raised levels of humidity amongst the boulders.

WHAT NEXT?

The unique environment of Black Mountain is home to at least two other animal species that exist nowhere else. In the 1970s, when the Rock Frog was first sighted, a gecko and a skink were also discovered. It is quite possible that there are other, as yet unknown, animals living among the boulders, awaiting discovery.

Appearance and Behaviour

The Rock Frog's back is yellow to light tan, sometimes with darker markings. Its black eyes and yellow body make it easy to see at night. The male calls from deep within crevices, calling more often during the day, perhaps because it lives in semi-darkness most of the time. Its call is a series of clicks made over several seconds. This frog is active at night when it forages on the surface of the rocks. The well-developed discs on its fingers and toes help it to clamber over rocks. Nothing is known of the breeding habits or life history of this frog, which is not surprising since they are so hard to find.

Water Frog

8 cm

The Water Frog probably arrived in Australia from New Guinea when the two countries were one.

Although true frogs — the frogs of the family Ranidae — are found all over the world, Australia has only one representative of this group. The Water Frog is also known as the Wood Frog or the Australian Bullfrog. Its distribution is limited to areas in northern Australia and it is normally found around the edges of permanent streams, ponds or lagoons where it remains in or near grass or dense vegetation. As it is also found in New Guinea, it probably migrated to Australia thousands of years ago when no sea separated the two countries.

Description

The Water Frog has a distinct fold of skin running from its eyes to its legs. It has long legs and an elongated body. This frog is a powerful leaper and strong swimmer. The male Water Frog is unique amongst Australian frogs in having two vocal sacs, one on each side of its head. All other Australian frogs with vocal sacs have a single sac under the chin. The Water Frog's call sounds like a series of quacks, similar to those of a duck, and it is usually heard throughout spring, summer and autumn. The Water Frog eats most animals that are small enough to cram into its mouth, including other frogs.

Up to several thousand eggs are laid in a large clump, but not in foam as some other species' are. Tadpoles grow up to 6 cm in length and are black and gold in colour.

NOT LONG ENOUGH

Although some true frog species are killed to provide frogs' legs for restaurants, fortunately for the Water Frog it doesn't have long enough legs to satisfy this peculiar food preference. For the time being at least it is safe from this particular group of human predators.

How do Frogs get Around?

*M*ost frogs jump, but some are better at jumping than others. The champion jumpers can leap a metre at a time, while those with shorter legs have to be content with making more of a bounce than a leap. They jump mainly to get away from predators or to move quickly towards their intended prey. Some frogs walk rather than hop and a few run or scuttle. Many frogs are good at climbing and spend considerable time off the ground, using their expanded finger and toe discs to help them stick to vertical or slippery surfaces.

All frogs can swim, although some use different strokes to others. The most common methods of swimming use either a dog paddle stroke, making use of all four limbs, or a 'frog kick', employing only the hind limbs.

Good jumpers can leap over a metre.

Do Frogs Have Parasites?

THANKS FOR HAVING ME
Many of the parasitic worms that infect frogs have a strange life cycle that involves several hosts. Insects are the first to be invaded by the worms, probably when they feed on the dung of an infected snake. The frog then eats the insect and the worms continue living inside the frog. Then, when the frog is caught and eaten by a snake, the parasitic worms take up residence in the snake's stomach and eventually lay their eggs within the host. The eggs are then passed out of the snake in its faeces where they are eaten by insects, and the life cycle continues.

A frog may act as host to several types of parasitic worms, including roundworms, tapeworms and flukes. Some worms attach themselves to tissue, while others move freely inside the body of the frog.

For some of these parasites their time spent living at the frog's expense is only part of their life cycle. Such parasites enter the frog's body inside prey that the frog has eaten and complete their development only if the frog is in turn eaten by a snake.

Frogs can also be hosts to mites and leeches. There is also a frog fly that lays its eggs in places frequented by frogs. The newly-hatched maggots seek out frogs and burrow under their skin where they live and move about slowly, apparently without affecting the frog.

A Blue Mountains Tree Frog with maggoty freeloaders.

How Can I Identify Frogs?

*T*he best way to find out what kind of a frog you have found is to consult a field guide. If you can get one that deals only with the frogs in your region or state, so much the better.

The main field guides for frogs are listed at the back of this book. These have colour photographs of each species together with a description of the frog. Most have what is known as a 'key', which is a series of statements about frogs. You decide which

A key will differentiate Peron's Tree Frog from Tyler's Tree Frog.

statement best fits your frog and then follow the directions to the next statement. By this method of elimination you finally arrive at the correct frog. Some species can be identified by the photographs alone, but many are very similar and an accurate identification can be made only by using the keys.

How Can I Recognise Different Frog Calls?

*A*nother way to identify a frog is by its call. There are a range of audio tapes of frog calls available, and these usually include a commentary as well.

Using a small, hand-held cassette recorder you can record a frog's call and then identify the frog that made it by comparing your recorded call with those on the identification tapes. With practice you will be able to identify the calls of different species with-

out needing to listen to the tapes each time.

These identification tapes are also useful if you want to find out what frogs are in a particular pond. Male frogs will often respond when they hear a recording of another male of the same species calling.

Some of these tapes are listed in the 'Further Reading' section at the back of this book.

Male frogs will often respond to frog-call recordings.

Where and When do I Find Frogs?

Dams are a good place to track down frogs.

*L*ook for frogs where you hear them calling. The frogs that you hear calling will be males and they will usually call close to or in a body of water. Most frogs call in spring and summer and are most active immediately after rain.

Farm dams, ponds, swampy areas and creeks are good places to find frogs. In arid regions, farm dams, flooded claypans, roadside ditches or any other flooded depression will probably have its quota of frogs.

Night is the best time to search for frogs as that is when they will be at their most active and calling loudest. Finding frogs during the day usually means looking for them in their resting places under rocks, logs and debris or up trees, although a few species are active at this time. You are unlikely to find a burrowing frog during the day as it will have dug in until nightfall. Any that are above ground are usually hiding in debris or in the water.

What Equipment do I Need?

*L*ooking for frogs inevitably means getting wet, so old clothing is essential, and lots of pockets will come in handy. Waders are useful in shallow water, but watch out for the deeper holes. The most essential piece of equipment is a waterproof torch with a powerful beam. Best of all are head torches that leave both hands free. If you are frogging at a remote site, then a second torch is advisable; at least carry a spare bulb and batteries.

The best containers for holding frogs are clear plastic bags that can be inflated and tied at the neck with a rubber band. While tadpoles need to be carried in water, don't put frogs in bags filled with water as the frogs will drown. The only other essential gear you'll probably need is mosquito repellent!

FOLLOWING THE LAW
Before you embark on a frog-hunting expedition check the regulations for your area with the wildlife or conservation authority in your State to find out what you can and cannot do. Also, if you are new to searching for frogs at night, remember that it is easy to lose your bearings in the dark. Always know which way you are facing in relation to the direction from which you entered the area.

Tadpoles are easily caught in nets.

How do I Catch a Frog?

*F*rogs are fairly easy to catch. They normally stay very still to avoid detection, only jumping or crawling away at the last moment. Experience will soon tell you when to make a grab to secure the frog. As a rule, frogs shelter out of the sun during the day.

Whether you are trying to catch frogs in the day or at night, the main problem is to get a hold on them. Because of their shape and because they are moist, frogs easily slide

Frogs are easier to catch with the help of others.

through your fingers or slip out of your closed hand. This is where a plastic bag comes in handy. Once you have the frog in your closed hand, put your hand inside a plastic bag, release your grip and then withdraw your hand. Remember to secure the bag with an elastic band. The frog is now safely inside the bag and can easily be viewed. This method is also less stressful for the frog.

How Can I Find Frogs in the Dark?

*Y*ou may know that an area is full of frogs because you can hear them calling, but in the middle of the night they can be hard to pinpoint. This is especially difficult as many will be well camouflaged or hiding in thick grass or leaf litter. There is an ingenious way of spotting your target, however — by triangulation.

All you need is three people with strong torches. Each person stands 5–6 metres from the next and faces the direction they hear the call. On signal, everyone shines their torch in the direction of the sound and where the torch beams cross is the general location of the frog. You will usually find it without too much searching.

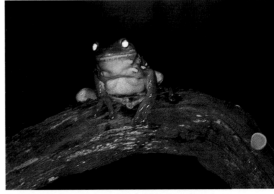

In torchlight, larger frogs can be seen by their eyeshine.

How do I Take Good Photographs of Frogs?

It is best to photograph frogs in their natural habitat.

*M*ore than anything else, you need patience to photograph frogs. Occasionally you'll find the perfect subject, sitting still in exactly the pose you want. The other 99 times your target will be jumping about or stubbornly staying in the wrong pose. Or it will move just as you click the shutter. Really, it is not much different from child or pet photography.

The best way to get good photographs of frogs is to have an assistant. This person can manipulate the animal into an appropriate pose, move it to a more suitable spot or keep it still while you click. It is better to photograph the frogs where they live. You can catch frogs and photograph them in another place, but this means you also have to collect some 'props', such as a branch, rock or plant from the original site and then return the frog and props to the site when you've finished.

What Photographic Equipment Should I Use?

*B*ecause frogs are small animals, the only really satisfactory equipment is a single lens reflex (SLR) camera with zoom or macro lenses. With this type of equipment the frog will fill most of the frame. It doesn't have to be an expensive camera, but it must be able to get small subjects in focus and filling the frame. The camera must also be able to achieve the correct exposure.

As photos will often have to be taken in poor conditions, particularly in the field, a flash unit is essential. Many photographers use two units mounted on either side of the camera to provide good shadowing without loss of detail. Other photographers find that a single unit gives satisfactory results. Don't be afraid to experiment until you find the combination that suits you.

A zoom is often handy because it gives a bit of distance between you and the frog, which makes the frog a little less 'jumpy'.

A typical camera set-up for photographing frogs.

Does Photography Stress Frogs?

Handling frogs should be done with extreme care.

*W*hen you are photographing frogs, keep in mind that you are dealing with wild creatures that are easily stressed. Keep the handling of a frog to a minimum and, if possible, ensure that your hands are wet, as a frog's skin is very sensitive.

The welfare of the animal should always be your first consideration. Because you will need to take several shots of a frog so that you will have a selection to choose from, you may have to hold the frog temporarily in a container. There are several rules to observe. Ensure the container is out of the sun, and don't put a lot of frogs in the same container, especially if they are different species or different sizes. Make sure the frog is kept moist and does not overheat while being photographed.

Remember that a frog should always be returned to the place where it was found.

What About the Future?

*F*rogs have been around for more than 160 million years, while humans have only been sharing their environment for about 2 million of those years.

During their long existence, frog populations have suffered massive declines and mass extinctions from natural causes. It would be tempting, therefore, to view their current decline, although due to human intervention, as just another natural event. However, humans have reached a level of knowledge and technological skill that would enable them to minimise their effect on other species — if they really wanted to.

A PIT tag used to identify individual frogs.

Pollution in all its forms is a very real problem and is the cause of many local extinctions. Habitat destruction is a worse problem, with wetlands drained for agriculture or other developments, and creeks and streams encased, for human convenience, in concrete pipes.

Conservation and wildlife authorities have so far been rather ineffectual since they cannot act on a large scale without the backing of politicians and the public. Without a major change in human values, the future for frogs is very bleak indeed. Although hampered by lack of funds, some research into the life histories of various frog species is being done. Some research even involves electronic tagging of the frog just beneath its skin with Passive Integrated Transponder (PIT) tags so individual frogs can be identified when next encountered.

A Checklist of Australian Frogs

Listed below are the names of the frog species described in this book. Their scientific names further indicate their relationships to one another. There are many other species of frog in Australia and a list of further reference material is provided at the end of this book for those who wish to find out more.

FAMILY HYLIDAE (TREE FROGS)

Blue Mountains Tree Frog	*Litoria citropa*
Broad-palmed Frog	*Litoria latopalmata*
Cave-dwelling Frog	*Litoria cavernicola*
Dainty Green Tree Frog	*Litoria gracilenta*
Desert Tree Frog	*Litoria rubella*
Eastern Dwarf Tree Frog	*Litoria fallax*
Green and Golden Bell Frog	*Litoria aurea*
Green Tree Frog	*Litoria caerulea*
Green-thighed Frog	*Litoria brevipalmata*
Magnificent Tree Frog	*Litoria splendida*
Tasmanian Tree Frog	*Litoria burrowsae*
Waterfall Frog	*Litoria nannotis*
Giant Frog	*Cyclorana australis*
Main's Frog	*Cyclorana maini*
Short-footed Frog	*Cyclorana brevipes*
Water-holding Frog	*Cyclorana platycephala*
Warty Water-holding Frog	*Cyclorana verrucosa*

FAMILY MYOBATRACHIDAE (GROUND FROGS)

Giant Banjo Frog	*Limnodynastes interioris*
Marbled Frog	*Limnodynastes convexiusculus*
Northern Bullfrog	*Limnodynastes terraereginae*
Ornate Burrowing Frog	*Limnodynastes ornatus*
Pobblebonk	*Limnodynastes dumerilii*
Salmon-striped Frog	*Limnodynastes salmini*
Spencer's Burrowing Frog	*Limnodynastes spenceri*
Spotted Grass Frog	*Limnodynastes tasmaniensis*
Striped Marsh Frog	*Limnodynastes peronii*
Carpenter Frog	*Megistolotis lignarius*
Giant Burrowing Frog	*Heleioporus australiacus*
Painted Frog	*Neobatrachus pictus*
Shoemaker Frog	*Neobatrachus sutor*
Sudell's Frog	*Neobatrachus sudelli*
Trilling Frog	*Neobatrachus centralis*
Great Barred Frog	*Mixophyes fasciolatus*
Crucifix Toad	*Notaden bennettii*
Desert Spadefoot Toad	*Notaden nichollsi*
Baw Baw Frog	*Philoria frosti*
Sphagnum Frog	*Kyarranus sphagnicolus*
Yellow-bellied Mountain Frog	*Kyarranus kundagungan*

Nicholls' Toadlet	*Metacrinia nichollsi*
Sharp-nosed Torrent Frog	*Taudactylus acutirostris*
Sunset Frog	*Spicospina flammocaerulea*
Brown Toadlet	*Pseudophryne bibronii*
Corroboree Frog	*Pseudophryne corroboree*
Red-crowned Toadlet	*Pseudophryne australis*
Common Eastern Froglet	*Crinia signifera*
Red-thighed Froglet	*Crinia georgiana*
Karri Frog	*Geocrinia rosea*
Smooth Toadlet	*Uperoleia laevigata*
Tanami Toadlet	*Uperoleia micromeles*
Fletcher's Frog	*Lechriodus fletcheri*
Hip-pocket Frog	*Assa darlingtoni*
Tusked Frog	*Adelotus brevis*
Southern Gastric Brooding Frog	*Rheobatrachus silus*
Sandhill Frog	*Arenophryne rotunda*
Turtle Frog	*Myobatrachus gouldii*

FAMILY MICROHYLIDAE (NARROW-MOUTHED FROGS)

Northern Territory Frog	*Sphenophryne adelphe*
Rain Frog	*Sphenophryne pluvialis*
Robust Frog	*Sphenophryne robusta*
Ornate Frog	*Cophixalus ornatus*
Rock Frog	*Cophixalus saxatilis*

FAMILY RANIDAE (TRUE FROGS)

| Water Frog | *Rana daemeli* |

FAMILY BUFONIDAE (TRUE TOADS)

| Cane Toad | *Bufo marinus* |

INDEX